Field Palaeontology

second edition

Roland Goldring

Longman

Pearson Education Limited
Edinburgh Gate
Harlow
Essex CM20 2JE
England
and Associated Companies throughout the world

Visit Addison Wesley Longman on the world wide web at:
http://www.awl-he.com

First published in 1991 as *Fossils in the Field*
This edition published 1999

ISBN 0 582 35625 3

British Library Cataloguing-in-Publication Data
A catalogue record for this book is available from the British Library

Library of Congress Cataloging-in-Publication Data
A catalog record for this book is available from the Library of Congress

Typeset by 35 in 9.5/11pt Times
Produced by Addison Wesley Longman Singapore (Pte) Ltd.,
Printed in Singapore

Contents

Preface

Every fossil in the laboratory, museum or on the mantlepiece was collected in the field.

The past decade has seen immense changes in palaeontology and in the study of sedimentary rocks in general. Undoubtedly, many of them have been initiated by the needs of the petroleum industry. Allostratigraphy (sequence stratigraphy) has revolutionized classical stratigraphy and imposed a basic framework for biostratigraphy. In turn, the need for greater refinement in stratigraphical correlation has led to high-resolution biostratigraphy.

In palaeontology itself, the field discoveries and advances in the understanding of soft tissue preservation have been at all stratigraphic levels. At the other end of the taphonomic spectrum, shell beds now provide a new range of information: ecological and stratigraphical. At the same time, advances continue in functional morphology and in systematic palaeontology. All go back to and depend on field observations.

The need to interpret ancient facies in as much detail as possible has led to an integrated approach in trace fossil studies. Attention is now focused on ichnofabrics, which combines classical sedimentology, sequence stratigraphy and ichnology. In the first edition trace fossils formed sections of several chapters. I have decided to give a separate chapter to trace fossils and bioturbation because, however one views them, their formation, taphonomy and applications are so utterly different from those of body fossils. It is only their naming that makes us suppose they are that closely related! Particularly exciting, as I write this, are the advances in vertebrate and insect ichnology, and their contribution to palaeocommunity studies.

Fossils comprise a unique data set, providing answers to a range of questions and having application to numerous theories. The aims of this book are therefore to provide a basis for evaluating the information potential of fossiliferous sediments, and then to give an outline of the strategy and tactics that can be adopted in the field. Some readers will be disappointed that the text does not provide a simple user-friendly list of jobs to be done at a site. This is because of the numerous information categories held in fossiliferous rocks; and because, even within a single facies, such as shallow-marine sandstones, the recoverable information is so variable (depending on the tectonic style, diagenetic history and type of weathering).

The text assumes that the reader will have a general knowledge of the plant and animal phyla. Chapter 4 discusses how to go about identifying body and trace fossils, the significance of misidentification and how to recognize pseudofossils. No keys for identification are included because they are not feasible at this level for vertebrate or plant fossils, and trials with undergraduate classes have shown that if students had enough knowledge to use the keys then they did not need them. Furthermore, keys cannot cope with many of the modes of preservation, e.g. a discrete spar-filled cephalopod camera. The reader will probably have carried out some mapping in the field area, or made a site investigation, and will be familiar with the geological timescale and basics of sedimentology. The book is not designed as a simple field handbook but as a guide, prompt and information source to be used at base, at outcrop and in the laboratory or core shed. All earth science students now have hands-on experience of core, emphasizing the broad and integrated approach to 'soft-rock' studies.

The book collates many facts, normally only available in reference collections and libraries, without recourse to which it may not be possible to pursue a particular line of investigation in the field. For example, approaches in field observation, largely influenced by geological theory, also draw on developments in other branches of science. In palaeontology it is the theoretical advances in biology and ecology that have made the most impact; also significant are advances in sedimentology, tectonics, geochemistry and biochemistry. The text thus aims to make the reader aware of relevant concepts from these rapidly developing disciplines. Furthermore, important as fieldwork is, many geological happenings affect a biota without leaving clear visual evidence at the exposure; anoxic events and iridium events are two examples. This means that information must be collated from elsewhere. One may be suspicious, one may speculate, and the answer may be forthcoming from later laboratory investigation of samples that are precisely located. Thus a good many facts are included in the text, but due to limitations of space, full documentation is not permissible.

Although this book is essentially concerned with fossils in the field, some of the most exciting and important finds in recent years have been made in our heritage collections: national museums, university and school collections, survey archives and private collections. In the past, as now, collectors may have been unsure of the identity of a particular specimen, or were puzzled by an odd association or type of preservation; or thought it odd that such and such a species should have been found where it was. It was not always prepared as expertly as it might be nowadays, but it was labelled and deposited. What a treasure chest these collections are, especially for palaeontologists, palaeobotanists and sedimentologists.

Acknowledgements

It was Rose Dixon (formerly of the Open University Press) who originally got the project under way. To Ian Francis and Lesley Evans (formerly of Longman) I am immensely grateful for their demanding but always constructive criticism and real interest. This edition owes much to Matthew Smith and his colleagues for their encouragement and advice, and especially to Bob Farmer for his editorial skills. Many colleagues and friends generously went over, smaller or larger sections for the first edition or made useful suggestions. All were acknowledged in the first edition and now again – thanks. Since publication colleagues, students and classes have contributed in more ways than I expect they realize. First thanks must got to Andy Taylor of Ichron Limited, who is leading the field in the applications of ichnology. Through him I have been able to examine and discuss core from many facies and many fields, made available with the generous assistance of many petroleum geologists and companies. To all warm thanks. I also thank Jordi de Gibert, Wang Guanzhong, Arpad Magyari, Jon Radley and Sören Jensen for discussion and enthusiasm. Alfred Uchman and Andres Wetzel kindly sent an advance copy of their paper.

I would also like to record my gratitude to Richard Bromley, John Pollard and Dolf and Edith Seilacher for their friendship and endless discussions over many years.

Figs 2.6 and 8.20 are based on or include unpublished figures by Robert Riding and John Pollard. The following gave material used for figures: Bill Baxendale (3.9), L. I. Salop (3.12), Gunter Freyer (3.19), Andy Hughes (5.8), David Stephenson (5.8), Alan Kendall (8.4), Mike Mayall (8.13), Stephen Eager (8.4), Andy Taylor (8.7, 8.29), Soren Jensen (8.11), George Best (8.13), John Evans (8.13), Jordi de Gibert (8.30), Ben Bland (8.30). John Allen kindly allowed the specimen in Fig. 4.3(d) to be photographed.

We are indebted to the following for permission to reproduce copyright material: the author, Dr D. J. Bottjer for Fig. 8.9 (Droser and Bottjer, 1990); Blackwell Science for Fig. 2.8 (Bridges and Chapman, 1988); Cambridge University Press for Figs 4.1 and 4.2; Schweizerbart'sche Verlagsbuchhandlung for Fig. 6.10 (Schäfer, 1969); Société Géologique de France and the author, Professor C. Pomerol for Fig. 7.1 (Pomerol and Cavelier, 1986); the Geological Society, London for Figs 8.8, 8.23, 8.24 and 8.28.

Every effort has been made to trace the owners of copyright material, but in a few cases this has proved impossible and we take this opportunity to offer our apologies to any copyright holders whose rights we may have unwittingly infringed.

Safety in the field

Forewarned is forearmed.

You are responsible for your own safety

Ascertain what essential equipment is required, and what are the safety procedures for the area to be visited. For north-west Europe, follow the rest of these guidelines.

Essential equipment

- Map and compass;
- watch;
- whistle;
- small torch;
- first-aid kit;
- rain and windproof gear, including a brightly coloured item;
- warm clothing reserve, including polythene survival blanket;
- sun hat;
- high-calorie food reserve;
- water container;
- walking boots: hard hat and goggles;
- knife.

General procedure

- Listen to the daily weather forecast (including wind direction), which may determine where it is prudent to work.
- Take account of the time and height of tides when planning coastal work.
- Write down each day your approximate route, working area and time of return, and leave it for others to see.
- In worsening conditions do not hesitate to turn back if it is still safe to do so.

If you get lost, disabled, benighted, or cut off by the tide, do not hesitate to stay where you are until conditions improve or until you are found. Supposed short cuts can be lethal.

Distress code

- *On mountains* 6 long blasts, flashes, shouts or waves in succession repeat at minute intervals

- *At sea*   signal SOS
  3 short then 3 long then 3 short blasts or flashes
  pause and then repeat

Rescuers reply with 3 blasts or flashes repeated at minute intervals

Further reading

Geologists' Association *A code for geological field work*. Geologists' Association, London

Geological Survey New South Wales *A code for geological field work*. Geological Survey NS Wales Mining Museum, Sydney

American Geological Institute (1992) *Planning for field safety*. AGI, Annapolis MD

Geological Society *Code of practice for geological visits to quarries, mines and caves*. Geological Society, London

Climber and Rambler *Mountain safety: basic precautions*. Climber and Rambler, London

Earth Science Teachers Association *Safety in earth science fieldwork*. ESTA, Department of Geology, Oxford Brookes Univeristy, Oxford

Committee of Heads of University Geoscience Departments *Safety in geoscience fieldwork: precautions, procedure and documentation*. Geological Society, London

British Mountaineering Council (1988) *Safety on mountains*. Manchester University, Manchester

References cited on inside back cover

Dunham, R. J. (1962) Classification of carbonate rocks according to depositional texture. In Ham, W. E. (ed.) *Classification of carbonate rocks*. Memoir of the American Association of Getroleum Geologists I, Tulsa OK, pp. 108–21

Embry, A. F. and Klovan, J. E. (1972) Absolute water depth limits of Late Devonian paleoecological zones. *Geologisch Rundschau* **61**:672–86

Folk, R. L. (1959) Practical petrographic classification of limestones. *Bulletin of the American Association of Petroleum Geologists* **43**:1–38

Folk, R. L. (1962) Spectral subdivision of limestone types. In Ham, W. E. (ed.) *Classification of carbonate rocks*. Memoir of the American Association of Petroleum Geologists I, Tulsa OK, pp. 62–84

Pettijohn, F. J., Potter, P. E. and Siever, R. (1973) *Sands and Sandstone*, Springer, Berlin

CHAPTER 1

Principles and classification

The best geologist is the one who has seen the most rocks.

H. H. Read and J. V. Watson
Beginning Geology

All geologists need to have an appreciation of fossils. The main object of this text is therefore to draw attention to the usefulness of fossils, as seen in the field, to individuals who will be looking at fossils with different objectives, and with different backgrounds and experience. This chapter discusses first the various categories of information that can be obtained from fossils, and then outlines a classification of fossiliferous sediments that reflects these information categories. The object is, at an early stage of any investigation, to assess the value of a site, and how the information potential may respond to vertical and lateral change.

A group of palaeontologists at a fossiliferous site pursue a number of different goals. Some are concerned with determining the biostratigraphy, some with the palaeoecology, others are searching for especially well-preserved material. Many follow carefully prepared sampling programmes, in order to collect fresh material for micropalaeontological or geochemical analysis. Similarly, the sedimentologist, having logged a section and determined the lithologies and current directions, has to consider the available biological data to make a convincing palaeoenvironmental analysis. It is to the palaeontologist, palaeoecologist and sedimentologist that this text is principally directed: how to recognize the potential of fossiliferous sediments and analyse the sediments to best advantage. The task of palaeoecologists and sedimentologists concerned with palaeoenvironmental interpretation is particularly demanding because they have to be at the same time palaeontologists, palaeoecologists, hydraulic sedimentologists, sedimentary geochemists and sedimentary mineralogists. The best way to deal with a particular site is to take each of these roles in turn, keeping in mind the way in which they integrate with the other.

The biggest hurdles to be faced in the organic world are the colossal diversity of life, today and in the past, and the unpredictability of evolution. Added to this is the problem of determining to which group of animal or plant a fossil or fossil fragment belongs, especially when only an impression is available, or it has been deformed tectonically.

If the immediate urge on locating a fossil is to extract it, then pause and consider what information might be forthcoming from a careful examination of its position and attitude in the sediment. A detective investigation is under way and there is seldom a simple answer and rarely can every line of information potential be completely followed through.

1.1 Categories of information

Systematic and morphological Information about the morphology and organization of hard and soft parts that can be used in identification and classification.

Physiological and ethological Information that can lead to an understanding of the function (physiology) and behaviour (ethology) of the organism preserved as a fossil. Good information is likely to be provided by material fossilized more or less in the position in which it lived (autochthonous), where soft tissue has been preserved, and by trace fossils.

Evolutionary Information that can lead to an appreciation of the evolutionary position and evolutionary pattern of the fossilized organism within a particular unit. The pertinent question is whether, in a sequence, it will be rewarding to sample sequentially to gain an appreciation of evolutionary changes. Stratigraphic breaks and facies changes need to be minimal.

Energetic   Information that can be used in assessing ancient productivity and ecological energetics. Preservation of growth laminae is important for any work in this field. (The word *energetics* means energy transformations within communities.)

Ecological   Information that can be used to determine ancient ecological relationships and palaeo-environments. Autochthonous material will be particularly useful, together with indications of trophic niche (such as predation), interspecific relationships, relationships to substrate surface, etc.

Hydraulic, stratinomic and younging   Information that can be used for determination of the hydraulic regimes, stratinomic aspects and as way-up, younging (geopetal) criteria. This information differs from preceding categories in that autochthony is not necessarily involved, and it is generally the hydraulic and stratinomic attributes of the fossils that are of interest. (The word *stratinomy* means the processes between death and final burial.)

Stratigraphic   Information that can be used for stratigraphic correlation and the identification of bounding surfaces. Here the degree of autochthony or allochthony, fragmentation and dissociation is less important providing there has not been significant reworking, and providing that identification is still possible. A fragment of a biostratigraphically important taxon that still retains its hallmark will be sufficient.

Diagenetic   Information that can be used to appreciate diagenesis and diagenetic history. Shape, structure and composition of fossils are the intrinsic variables that relate to the chemical and compactional changes in sediments, which largely follow completion of stratinomic processes and precede later diagenetic change associated with tectonic history. (The word *diagenesis* means the changes – chemical, physical and biological – that occur after initial deposition.)

Besides these categories there will be information on *geotechnique* (distribution of porosity and permeability attributable to skeletal morphology and to

bioturbation) and implications for reservoir characterization and management.

1.2   Principles

Two principles, in particular, are involved when investigating fossiliferous sediments:

1   Different groups of fossils have different information potential, either because of their inherent attributes, or because of taphonomic considerations.
2   Different types of fossiliferous sediment yield different categories of geological information.

The first principle is one that is generally appreciated. Systematic and stratigraphic classifications of fossils are used at an early stage of geological experience. Compare the relatively small amount of information that can be obtained from the morphology of a fossil gastropod in contrast to the much greater amount that is generally available from an echinoid test. Poor preservation may further reduce the information from a strongly recrystallized gastropod shell. Or compare the high stratigraphic value of many graptolites with the low value of contemporary bivalve molluscs. Similarly, it is shells with moderate convexity and bilateral symmetry that are the most useful types in the determination of flow characteristics. Also, benthonic calcareous algae are more useful in environmental interpretation than are many other groups.

The second principle may be understood by scanning Fig. 1.1 or, more closely, a section in Upper Jurassic (Oxfordian) sands and limestones (Figs 1.2 and 1.3 to 1.5; see also Box 1.1). In the basal sands, the narrow shafts and local galleries (*Ophiomorpha nodosa*) made by burrowing arthropods are useful as indicators of an estuarine environment; whereas more precise stratigraphical information may be obtained from thin muddy partings, yielding spores, pollen and occasional microplankton. The palynofacies supports interpretation of the sands and muds as representing an estuarine depositional environment. The only moderately good stratigraphic information from the upper, calcareous part of the section is from relatively uncommon and poorly preserved ammonites from the oolites. No ammonites have yet been discovered from the coralliferous part of the section and its age is somewhat uncertain. Bivalves at the coarse-grained, shelly base to the transgressive carbonates often show preferred orientation and provide information

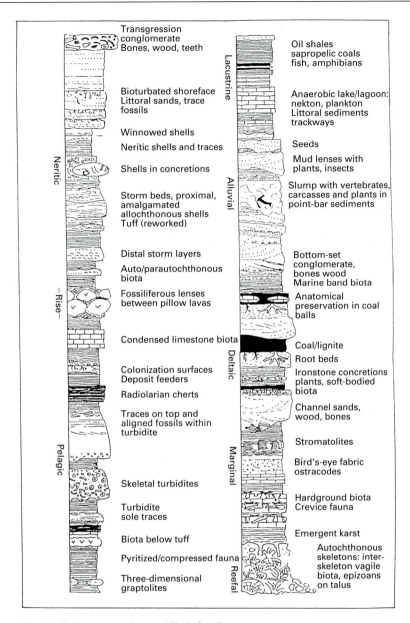

Fig. 1.1 Facies succession and likely fossil ocurrences.

Fig. 1.2 (a) View of quarry face, now an SSSI, at Stanford Quarry, Stanford-in-the-Vale, Upper Jurassic. Jozef Kazmierczak and Esther Jamieson are standing on the sequence boundary between estuarine point-bar sands below and limestones above. They are examining the base of the coralliferous unit. (b) Print from a thin section of *Fungiastrea arachnoides* showing diagenetically enhanced zones of thick septa, and light zones with poor preservation. (Scale bar = 5 mm)

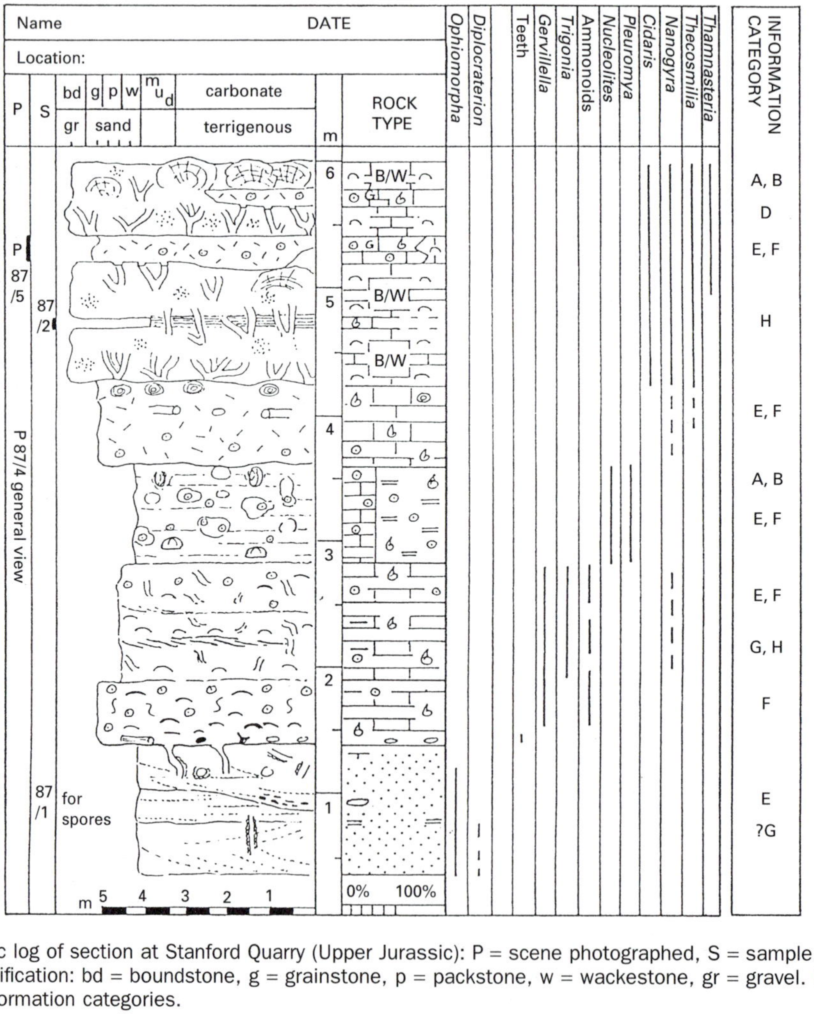

Fig. 1.3 Graphic log of section at Stanford Quarry (Upper Jurassic): P = scene photographed, S = sample, with number. Sediment classification: bd = boundstone, g = grainstone, p = packstone, w = wackestone, gr = gravel. Letters A to H indicate the information categories.

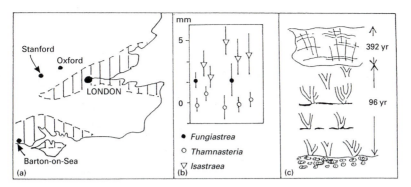

Fig. 1.4 (a) Location of quarry at Stanford-in-the-Vale, central England, along with Tertiary basins and site of Barton-on-Sea (Box 5.1). (b) Growth rates of three corals in the Corallian with mean and standard deviation, determined from measurements of diagenetically enhanced zones. (c) Generalized representation of section to illustrate coral growth rate and minimum time of formation. Part (b) adapted from Ali (1984).

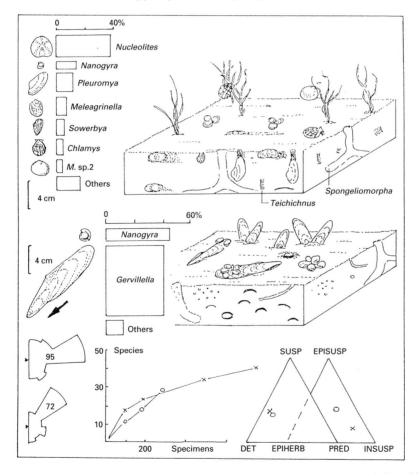

Fig. 1.5 Reconstruction of two associations at Stanford Quarry: (1) *Nucleolites scutatus* association with its (skeletal) trophic nucleus (the species that account for 80% of the fossils). This is how Fürsich would reconstruct the nodular, muddy oolite 3 m above base; (2) *Gervillella aviculoides* association, corresponding to the lower carbonate beds, with trophic nucleus and orientations of *Gervillella* measured on fallen blocks, with number of specimens. Also shown are rarefaction (sampling) curves for the two associations (×) = *Nucleolites* and (○) = *Gervillella*; they plot the number of species against the number of specimens. Beside them are triangular diagrams for feeding habits (left) and substrate niches (right): SUSP = suspension, EPISUSP = epifaunal suspension, DET = detritus, EPIHERB = epifaunal herbivore, PRED = predator, INSUSP = infaunal suspension (for details, see Fig. 5.24); (×, ○) are the same as for the rarefaction curves. Adapted from Fürsich (1977).

Box 1.1 Mesozoic intertidal sands, neritic limestones and a coral bank

Stanford Quarry (Figs 1.2 and 1.3 to 1.5), a few kilometres to the west of Oxford, England, was an extensive shallow quarry, formerly worked for the fine-grained sand which is of good mortar quality. A face in the carbonate sediments has been conserved as a Site of Special Scientific Interest (SSSI). A working quarry to the south better displays the lower units. The strata, of Upper Jurassic age, dip gently southwards. The objectives of the investigation were to carry out a sedimentological and palaeoecological investigation of the sands, limestones and coralliferous unit. The coralliferous unit is the only good exposure of such facies in southern England. An appreciation of the depositional environment of the sands would aid in their further exploitation. A reconnaissance survey, lasting at least an hour, showed that there was extensive lateral change, which would need careful investigation. (As part of this initial investigation, estimates were made of the materials needed for making peels from the loose sand.)

The main site investigation was tackled in two stages. Firstly, the sands were mapped in detail at scales of 1 : 10 or 1 : 20, and the facies dissected with a trowel, and directions of foresets measured and recorded. The sands dip more steeply than the overlying limestones, which truncate the primary stratification. Mapping and logging several of the faces showed that the sands must have formed on an extensive point bar migrating southwards. Slender J-form shafts of *Ophiomorpha nodosa* dominate. Since *Ophiomorpha* is a marine indicator and clearly autochthonous, this means the section was part of an estuary (delta distributary), and it would be reasonable, as a first model, to extend working of the sands towards the north-west. More detail could be observed by searching for wind-etched surfaces. Sections were photographed in colour and in black and white. Fresh samples (100 g) of the muddy laminae were collected from identified horizons for stratigraphic and palynofacies analyses. Peels were made of selected areas, and samples taken for grain size analysis and clay mineralogy.

In the second stage of the investigation, the limestones were mapped and logged, paying particular attention to the nature of bed contacts. Traced laterally, a slightly irregular erosion beneath the coralliferous unit could be detected. The lenticularity of the shelly limestones (bio-oosparites) within the coralliferous unit was also investigated. A programme of alternating field and laboratory investigation concluded the study. This allowed some elements of the fauna to be identified in the field from British Museum handbooks, but most were identified in the laboratory from monographs. At this time, oriented samples (minimum dimension 10 cm) were collected for slabbing and thin-sectioning or acetate peel analysis; the position of each sample was recorded on the logs.

Examination of the coralliferous unit took considerable time. The phaceloid corals are mostly preserved as moulds in a cemented wackestone matrix. Their morphology was sketched, and blocks of the phaceloid and massive corals collected for examination and sectioning. Later these were used for a study (Ali, 1984) of coral growth and productivity (Section 5.2.2), a study which Insalaco (1996) has extended to Europe. Fresh samples of the clayey seams were collected for X-ray fluorescence (XRF) analysis and bags of the weathered clays collected with a knife and fine trowel for wet sieving. These were found to have a rich microbiota, including photo-negative, cryptic thecideidine brachiopods (Fig. 5.3), small gastropods and brittlestar elements, which would be included in the reconstruction of the unit. Specimens in regional and national collections were later examined, and discussions led to closer identification and better knowledge of the existing literature.

The coralliferous section represents a bank (with minimal relief) rather than a reef (with appreciable relief). A similar approach can be used for any largely autochthonous accumulation. The bedded limestones can also be matched throughout the Phanerozoic.

Sources: Ali (1984), Insalaco (1996)

---

on the nature of local erosional environments. The coralliferous unit, in which the branching coral *Thecosmilia* dominates, has an abundant and well-preserved associated biota, including regular echinoids and bivalves. Examination of the wackestone matrix reveals small gastropods, brittlestar 'vertebrae' and small thecideidine brachiopods (Fig. 5.3). The branching and massive corals have provided opportunity for growth and productivity studies. The muddy oolite also displays an autochthonous/parautochthonous biota of infaunal bivalves and the irregular echinoid *Nucleolites scutatus*. In this sequence the information potential changes rapidly corresponding with the facies changes.

1.3 A classification of fossiliferous sediments

Wilhelm Schäfer (1972) proposed a classification of marine sediments, applicable to ancient fossiliferous sedimentary environments. Evolving from a similar ecological work by Hermann Schmidt, the classification showed that stratification and completeness of the sedimentary record (for benthic organisms and sediments alike) were closely related to the degree of oxygenation. Schäfer's scheme (slightly modified in Fig. 5.16) is extended to 13 broad types of fossiliferous sediment (Table 1.1) to identify more closely the categories of information potential. Of primary concern is whether or not benthic organisms and traces are present; since their presence, if more or less in life position, normally indicates at least a degree of oxygenation at the substrate. Dominance of nektonic and/or planktonic fossils will generally suggest accumulation under anoxic or dysaerobic conditions, though this may not be readily resolved. In the fossil record, cephalopods, graptolites and pelagic trilobites are some of the most useful stratigraphic markers. In the Cretaceous chalk the sediment 'rain' of coccoliths and planktonic foraminifera is of stratigraphic importance. The chalk substrate was also for the most part richly colonized, as evidenced by the high degree of bioturbation and skeletal remains. Fossils of subaerial plants are mostly concentrated in facies quite distinct from animal fossils. The primary, if empirical, criterion of the classification is whether the site, or sedimentary unit, yields dominantly benthonic, dominantly nektonic and/or planktonic, or dominantly plant fossils.

Brett and Baird (1986) have coined the useful term *taphofacies* (Section 3.1) for sedimentary rocks characterized by particular, sometimes unique, combinations of preservational features, the result of similar taphonomic pathways. However, in their discussion of trilobite taphofacies, Brett and Baird were naturally most concerned with taphonomic pathways, the sorting, reworking and damage that may occur before final burial, rather than the early diagenetic, mainly chemical, aspects. In practice, stratinomic and diagenetic aspects are analysed separately at outcrop and the stratinomic aspects cannot be separated from hydraulic considerations. How stratinomy and early diagenesis reflect intrinsic factors, such as skeletal mineralogy, dissociation, disarticulation and fragmentation, as well as extrinsic factors, such as sedimentation rate and burial processes, are discussed more fully in Chapter 3.

The second criterion in the classification is the influence of taphonomic processes on the facies. Later diagenesis and tectonic history are too variable to be predictable from the depositional environment, although some sedimentary facies are more prone than others to dedolomitization or particular styles of tectonic deformation. Information potential is also influenced by the degree of stratigraphic condensation or expansion. Hardgrounds are by definition condensed, and pelagic shales and pelagic nodular limestones are also typically so; slumps tend to be expanded and associated with steep depositional gradients. But the organization of shelly fossils in turbidites is generally similar to their organization in shallow-water storm-generated sediments, such that the information potential is alike in most respects:

- Benthonic fossils predominate in types 1–6.
- There are two main types of fossiliferous deposit where pelagic or epiplanktonic fossils are the principal fossils (7, 8).
- Five types of fossiliferous deposit where subaerial plants are the principal fossils (9–13).

This classification cannot be exhaustive; there are intermediates and there are other types of skeletal accumulation associated with biogenic stratification and biogenic graded bedding described from modern environments (Section 2.3). Although it is useful to distinguish subaerial (mainly vascular) plants in volcanic associations, it is less important to separate skeletal accumulations in such facies for their information potential. Furthermore:

- Each type of deposit may be dominant in a sedimentary sequence, or may be present as only a minor facies. For instance, bituminous shales (7) may be a minor facies in a sequence dominated by rather unfossiliferous turbiditic muds and silts.
- Any of the types distinguished may contain fossils of such palaeontological significance, either because of the concentration of fossils or because of unusual tissue preservation, that a fossil-ore (Fossil-Lagerstätte) is represented (Section 5.5).
- Many of the fossils that are most typical of one or other type of fossiliferous sediment may be found in another. For example, anatomically preserved plants occasionally occur in marine shales.

Table 1.1 Thirteen categories of fossiliferous sediment

TYPE 1

Autochthonous buildups ranging from stromatolites to small bryozoan mounds to oyster beds to barrier reefs, where biogenic processes are dominant (Box 1.1)

(A, B) *SMPE*: good to excellent for many structures, but often loss of aragonite and/or dolomitization; potential information on astogeny and life attitude, though fossils may be difficult to release from matrix
(C) *Evolutionary*: poor
(D) *Energetic*: potentially good, e.g. corals, oysters, depending on amount of diagenetic alteration
(E) *Ecological*: potentially good, especially for interrelationships of biota
(F) *Hydraulic and stratinomic*: growth form may be indicative; also, talus within buildup and around margin may provide important information
(G) *Stratigraphic*: can be poor; age relationships often best established from underlying and overlying strata
(H) *Diagenetic*: complex early diagenesis ± meteoric, vadose or phreatic cements, dissolution, replacement; dolomitization common

TYPE 2

Well-stratified sediments (siliciclastic or calcareous), rapidly deposited, with hydraulic processes and products dominant. Ranging from fluvial sands, bioclastic shelf limestones and sandstones, storm beds and tempestites to bioclastic turbidites and fossiliferous tuffs

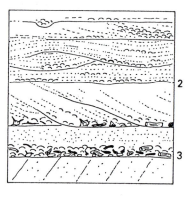

(A, B) *SMPE*: good to excellent, but morphological information may have to be pieced together from dissociated and broken material; preservation can be good, though is often mouldic
(C) *Evolutionary*: with high abundance of fossils, potentially good, but size sorting typical, or sediment washed
(D) *Energetic*: poor or difficult to resolve, because of sorting and mixing (time averaging)
(E) *Ecological*: mainly refers to environment(s) of provenance
(F) *Hydraulic and stratinomic*: generally informative from sedimentary structures and skeletal fabrics
(G) *Stratigraphic*: variable and often difficult to resolve because of reworking and introduction of derived material
(H) *Diagenetic*: in siliceous successions where sedimentation rapid ± early carbonate dissolution leading to moulds/replacements; compaction low; in calcareous successions early cementation often leads to good preservation; in regressive successions meteoric cements likely

TYPE 3

Lag concentrates at unconformities and discontinuities, with hydraulic processes dominant. Can also include fissure fills

(A–D) As for type 2
(E) *Ecological*: more difficult to resolve than in type 2 because of mixing, sorting and damage
(F) *Hydraulic and stratinomic*: generally distinctive
(G, H) As for type 2

TYPE 4

Mass flow, debris flow and talus deposits generally representing transport from shallow to deep water, from reefs or associated with submarine fault scarps. The biota was rapidly buried and records elements often absent or poorly preserved in the original environment. Examples include the Burgess Shale (Middle Cambrian, British Columbia) and Cowhead Breccia (Ordovician, Newfoundland)

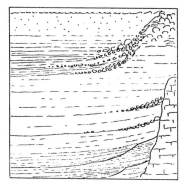

Table 1.1 (cont.)

(A, B) *Systematic and morphological*: good to excellent, and may include preservation of soft tissues

(C) *Evolutionary*: often very important source of information

(D, E) *Energetic and ecological*: difficult to determine because of redistribution

(F) *Hydraulic and stratinomic*: clear from sedimentary structures

(G) *Stratigraphic*: generally good but mixing can pose problems; clasts may be older or younger than matrix

(H) *Diagenesis*: likely to be complex, possibly involving predisplacement aspects

TYPE 5

Internally poorly organized (generally bioturbated, often heterolithic) sandstones and sandy mudstones, mudstones, limestones (packstones to wackestones), also some micritic limestones (most coccolithic chalks). Typically marine shelf sediments (also lagoonal and lacustrine) formed under mixed biogenic and hydraulic processes. Abundance of skeletal fossils influenced by net sedimentation rate so that 'end' members are shelly and bone gravels (starved of sediment) or poorly fossiliferous mudstones (starved of fossils).

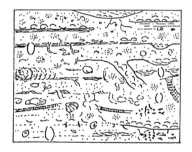

(A, B) *SMPE*: good to excellent for body and trace fossils, but disturbance or breakage by bioturbation likely

(C) *Evolutionary*: potentially good, but may be difficult to obtain large samples of a taxon

(D) *Energetic*: potentially good, but requires careful analysis

(E) *Ecological*: potentially good, with autochthony/ parautochthony common

(F) *Hydraulic and stratinomic*: can be difficult to establish, especially if primary structures destroyed by bioturbation, though way-up generally readily determined

(G) *Stratigraphic*: generally good to excellent from macro- and microbiota

(H) *Diagenetic*: early phosphatic, sideritic, pyritic concretions typical; early cements tend to be patchy; compaction high in muddy sediment ± preservation of aragonite

TYPE 6

Hardgrounds of various types and depicted here by planed Jurassic shallow-water hardground, and irregular Cretaceous Chalk omission hardground

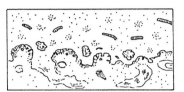

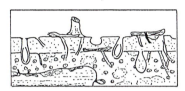

(A, B) *SMPE*: can be good for encrusting and boring biota, especially in crevices and within burrows/borings (coelobites); within hardground preservation good (e.g. preservation of aragonitic biota) when cements early

(C) *Evolutionary*: generally not applicable

(D) *Energetic*: difficult to interpret because of high loss of information

(E) *Ecological*: specialized but good, especially for succession of borers and encrusters

(F) *Hydraulic and stratinomic*: variable, but succession of colonizers may provide clues, e.g. truncated borings

(G) *Stratigraphic*: care required because of incomplete sedimentary record

(H) *Diagenetic*: compaction minimal, very variable cements especially in planar hardgrounds, where meteoric processes likely

TYPE 7

Well-laminted and often bituminous mudrocks, thin-bedded limestones (Plattenkalk), also radiolarian cherts and diatomites

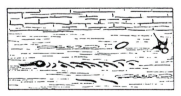

(A, B) *SMPE*: often exceptionally good to extraordinary

(C) *Evolutionary*: potentially excellent, but compression may hinder evaluation

(D) *Energetic*: not applicable because generally little or no benthos

(E) *Ecological*: generally good

(F) *Hydraulic and stratinomic*: seldom applicable, but if preferred orientation or imbrication present, then

Table 1.1 (cont'd)

substrate not continually anaerobic; way-up evidence often absent; minimal stratinomic disturbance

(G)   *Stratigraphic*: generally very good from micro- and macrobiota

(H)   *Diagenesis*: ± high compaction, pyritization (disseminated to nodular), skeletal dissolution common ± preservation of organic matter (including soft tissue)

TYPE 8

Nodular micritic limestones, the most typical being those packed with cephalopods (Ammonitico Rosso, Cephalopodenkalk)

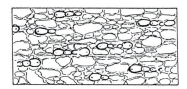

(A, B, C)   *SMPE and evolutionary*: often difficult to determine because of early dissolution of aragonite and calcite

(D)   *Energetic*: not applicable

(E, F)   *Ecological, hydraulic and stratinomic*: needs careful analysis because of possible reworking and condensation

(G)   *Stratigraphic*: generally good, allowing for preservational problems

(H)   *Diagenetic*: variable early diagenesis depending on water depth, i.e. ± within photic and aerobic zones; early cements (marine) common ± skeletal dissolution

TYPE 9

Peats, lignites and coals; coal balls and other indurated peats are particularly important. These are the plant equivalents to type 1

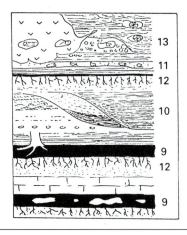

(A, B)   *SMPE*: generally poor (except for pollen, spores and in concretions); arthropod cuticle may be common, though often fragmentary

(C)   *Evolutionary*: good for microbiota

(D)   *Energetic*: potentially good if rate of accumulation can be determined

(E)   *Ecological*: distinctive but small-scale variation likely

(F)   *Hydraulic*: generally not applicable

(G)   *Stratigraphic*: generally poor (except from pollen, spores and in concretions)

TYPE 10

Well-stratified (often laminated) fluvial or lacustrine mudstones (e.g. Mazon Creek) siltstones and limestones (e.g. Green River, Eocene, USA), abandoned channel fills and point-bar sediments. Early diagenetic concretions important. Abundance of material makes them the plant equivalent of type 2

(A, B, C)   *SMPE and evolutionary*: generally good for leaves as compressions of impressions

(D)   *Energetic*: difficult to determine

(E)   *Ecological*: useful, but mixing may have occurred

(F)   *Hydraulic*: important to determine for ecology

(G)   *Stratigraphic*: generally useful from microflora

TYPE 11

Terrestrial slump deposits, often massive mudstones

(A–G)   As for type 10 but material generally less compressed

TYPE 12

Root beds and autochthonous stumps, fossil forests

(A–D)   Poor in root beds; variable in fossil forests, depending on mode of preservation

(E)   *Ecological*: good

(F)   *Hydraulic*: not applicable

(G)   *Stratigraphic*: spores and pollen often oxidized in fossil soil

TYPE 13

Volcanigenic associations: lavas, tuffs and volcanically derived sediments with calcification in basaltic terranes and silicification in rhyolitic terranes, e.g. Midland Valley (Carboniferous, Scotland), Florissant Formation (Miocene, Colorado). Several preservation modes are likely; concretions are most important. Note also volcanic vents, and associated mudflows of type 11

(A)   *Systematic*: can be optimal depending on degree of degradation

(B–E)   *MPEE*: association of plant with fauna

(F)   *Hydraulic*: not applicable

(G)   *Stratigraphic*: can be very important

SMPE = systematic, morphological, physiological and ethological.
MPEE = morphological, physiological, ethological and evolutionary.

1.4 The next stage

Following recognition of the different categories of fossiliferous sediment, consideration should be given to the investigation in progress. Both the time available and the amount of detail required will place limitations upon the course of study. Depending upon what information is of most concern, each category will require a rather different approach:

- If the only significant results are likely to be of a hydraulic nature, then emphasis should be placed upon the sedimentological aspects of the site (Chapter 6).
- Where contributions to knowledge of the morphology and functional morphology of a certain taxa may be determined, it may be more prudent to concentrate the field study on particular beds or intervals of the section under examination. Attention should also be given to the sedimentological setting and associated biota.
- In instances where environmental interpretation is the primary objective, the ecology of each taxa (autecology) should be determined initially, with further thought being given to hydraulic aspects.

A sophisticated study of palaeoenergetics would require a carefully prepared programme, and is beyond the scope of this book.

References

Ali, O. E. (1984) Sclerochronology and carbonate production in some Upper Jurassic reef corals. *Palaeontology* **27**: 537–48

Brett, C. E. and Baird, G. C. (1986) Comparative taphonomy: a key to paleoenvironmental interpretation based on fossil preservation. *Palaios* **1**: 207–27

Fürsich, F. T. (1977) Corallian (Upper Jurassic) marine benthic associations from England and Normandy. *Palaeontology* **20**: 337–85

Insalaco, E. (1996) The use of late Jurassic coral growth bands as palaeoenvironmental indicators. *Palaeontology* **39**: 413–31

Schäfer, W. (1972) *Ecology and palaeoecology of marine environments*. University of Chicago Press, Chicago IL

CHAPTER 2

Field strategies

If you know what it is you are searching for then you will find it.

The aim of this chapter is to discuss essential field strategies. Tools, techniques and tactics are summarized in Appendix A, with additional information in Chapter 4. Assessing the literature and making a reconnaissance are often tedious but always well worth the trouble, especially since they enable one to formulate more clearly the objectives of the field study, and to plan the most economical use of the time available in the field. Perhaps the most important 'rule' to observe is *not to hammer*, but firstly to search faces and soles carefully.

A section on stratification is included here because the *bed* is the basic sedimentary unit, and any fossil that is found *in situ* has to be related to the bed and its lower and upper surfaces. Changes at bed contacts reflect change in sedimentary environment and become the most conspicuous and, indeed, most important lines on the sedimentary log. An appreciation of bedding is therefore essential in logging outcrops and cores.

Most specimens collected in the field are, in reality, samples that may be subsequently used to illustrate an aspect of the facies or biota. If the aims of any investigation are carefully defined, then a sampling plan can be formulated. The field strategy that is required for buildups is rather dif-ferent from that required for stratiform fossiliferous occurrences. For both there is no short cut to the basic mapping of the site, and accurately recording the exact location of observations and specimens collected.

Basic field mapping and surveying techniques are covered by Barnes (1995), Butler and Bell (1988), Compton (1985) and Moseley (1981); details of all these texts can be found in the further reading at the end of the chapter.

What is particularly important is to have a good understanding of whole animal and plant biology and ecology, and the distribution and character of modern sedimentary environments. An account of sedimentary environments and facies is beyond the scope of this volume and reference must be made to the texts listed at the end of this chapter. Fig. 2.1 is included to show the distribution of the main sedimentary environments in simplified form and is amplified in Figs 8.8 and 8.18.

2.1 Previous work

Although there is something to be said for going into the field with an open mind, unprejudiced by the findings of previous workers, afterwards most of us generally regret having done so. The objective in making a literature survey is to establish the 'state of the art', the distribution and age of the rocks, the structure of the area and current problems. One should get as good an appreciation as possible of the biota likely to be encountered, so that the taxa can be readily recognized.

Begin with a regional textbook and follow with more detailed reports, surveys, memoirs and monographs. There is obviously a limit to the amount of literature that can be taken into the field, but glean information on locality lists and note preservation modes from the illustrations. It may be necessary to refer to books or papers borrowed from distant libraries, and some may need translation. Inspection of museum collections can prove valuable. Even more useful is to discuss plans with someone already familiar with the area. These aspects can take a lot of time, hence the need for ample preparation. When in the field contact local museums and geologists. This is also the time to assemble the necessary tools and

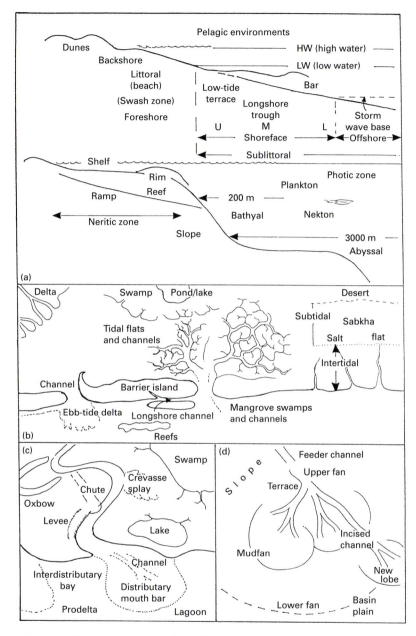

Fig. 2.1 (a) Horizontal and vertical zonation of sea and ocean; (b) shoreline environments with temperate and tropical (wet and arid), intertidal geomorphologies; (c) Coal Measure environments. Adapted from Scott (1979); (d) turbidite environments.

materials and to test them out fully. Delivery times may be several weeks or longer. For example, is it possible to transport peel-making materials (hazardous substances)?

Allow for the possibility that fossil names in older literature may have been subsequently revised, and

also that illustrations in prephotography days are not always as accurate as might be expected. Be prepared for cases of misidentification in older fossil lists and do not expect that collecting sites of a century ago will still be available. With older literature note that many new fields have since been researched. Faunal lists in a century-old paper may be detailed, but it is unlikely there will be substantial reference to the palaeoecology or sedimentology.

Be sceptical of claims in the literature that sediments are unfossiliferous, as they may abound with trace fossils, or fossils may be found once the nature of their preservation is understood, e.g. soft-bodied organisms in late Precambrian phosphatic concretions. At most, 'unfossiliferous' only means that no fossils have yet been found.

2.2   Reconnaissance

The object of a reconnaissance is to gain an appreciation of an outcrop or area sufficient to know where the best sections are situated, their limits and weathering profiles, the general accessibility and local problems (including obtaining permission to enter land), what tools, recording and photographic equipment are required and whether or not special techniques, such as peel making, are likely to be needed. Also get a good idea of the time required to carry out the objectives.

If good loose material is spotted, collect or flag it, since it may not be readily relocated, but do not spend time on detailed investigations. Nevertheless, this is the time to carry out carefully considered pilot studies. When the reconnaissance is completed reassess the objectives and the feasibility of completing them, and formulate as detailed a programme of work as possible, including a list of the jobs required, putting them in order of priority.

2.3   Stratification, bedding and cyclic sedimentation

The bed is the basic sedimentary unit. It can be recognized in most sedimentary facies, including many autochthonous buildups. Bedding is generally much less obvious in core. Fig. 2.2 attempts to portray in two dimensions the principal types of bedding that are generally associated with body and trace fossils. Note how difficult it may be to fix the lower as well as the upper boundary of many types of bed.

The term bed is frequently applied to sedimentary units greater than 10 mm in thickness, below which the term lamina is applied. This is an arbitrary definition. Lamination is present within a bed and results

from the physical processes operating during the formation of a bed. Do not refer to a thickness of sediment with lenticular or heterolithic bedding as a bed, because each alternation was formed under a different sedimentary regime. But a unit of climbing ripple lamina-tion may constitute a bed. For different thicknesses of bed, the most convenient method of de-signation is as mm bedding, or mm–cm, cm, dm, etc. Also consider carefully before using bed for an interval of mudrock. The term *event bed* is commonly used for a storm, turbiditic unit, or tempestite marking a single rapidly deposited sedimentary unit (Section 6.3). There are several types of bed.

Turbidites   Trace fossils on the sole of a turbidite event bed (deposited from a turbidity current) indicate preturbidite and deeply penetrating, post-turbidite burrowing activity (Fig. 8.22). Trace fossils on the upper surface indicate the shallow-penetrating post-turbidite suite possibly introduced with the turbidite flow. Within turbidites, which can be composite, body fossils tend to be scattered, but with preferred orientation and often convex surface down. Shells may be relatively uncompressed.

Interturbidite mudrock   Interturbidite mudrock is generally sparsely fossiliferous but it can yield important material. Water depth is reflected in preservation of calcareous skeletons (Section 5.3.2). Release of fossils often depends on cleavage relationships and weathering.

Storm beds and tuffs   Trace fossils on soles and upper surfaces are less strongly differentiated compared with turbidites. Body fossils occur in lags, concentrated in gutter casts, or in lenses associated with cross-stratification (including hummocky and swaley cross-stratification with, respectively, convex-up and convex-down intrabed discontinuities). Fossil fragments are often scattered through a bed. Amalgamation is common, and each unit (bed) may be associated with a different biota. Rarely, event beds may cover (smother) a biota.

Graded muddy sandstones or siltstones   Graded muddy sandstones or siltstones (marine shelf) are generally poorly fossiliferous.

Heterolithic stratification   Heterolithic stratification describes alternating mud and sand grade sediment; it is a term often applied to irregular

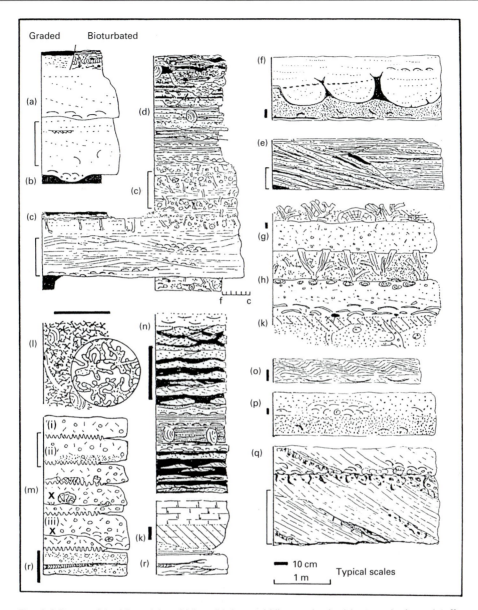

Fig. 2.2 Types of bedding: (a) turbidite; (b) interturbidite mudrock; (c) storm beds and tuffs; (d) heterolithic stratification; (e) inclined heterolithic stratification; (f) load-cast bedding; (g) event bed; (h) firmground and hardground; (k) underbed; (l) growth bedding; (m) pseudobedding – (i) unverifiable pseudobedding, (ii) exaggerated true bedding, (iii) verifiable pseudobedding; (n) cross-lamination; (o) ripple-drift lamination; (p) cluster of shells; (q) cross-stratification; (r) false indications of bedding.

alternations of mm–cm thickness. Where weathering allows, sandstone soles are often rich in trace fossils, e.g. *Cruziana*. Look for shelly partings, lenses and gutters. The shelly biota is often rich and diverse and often autochthonous or parautochthonous.

Inclined heterolithic stratification on point bars For palaeobiological purposes, this can be included with heterolithic stratification. The laterally accreted alternations of thin mudstones and siltstones of fluviatile and estuarine facies are often rich in plant fossils. Channelwards, packets of muddy sediment, the result of channel switching, are also a main source of plant fossils.

Load-cast bedding   Load-cast bedding is generally poorly fossiliferous, often with evidence for re-activation when in shallow-water facies, especially estuarine facies. If present, shells are distributed as in heterolithic stratification.

Convolute lamination   Convolute lamination is generally poorly fossiliferous.

Palaeosols   Palaeosols are formed under various climatic conditions; they mark floodplain deposits or, if associated with a marine succession, emersion events where they may be useful markers. A molluscan fauna may be present that was associated with the soil, or introduced with subsequent submersion.

Event bed   An event bed of bioclastic grains may truncate and smother coral growth.

Firmgrounds and hardgrounds   Firmgrounds and hardgrounds are associated with omission in sedimentation and recognized through colonization by encrusters and borers or burrowers. A mineralized crust is typical. As depicted in Fig. 2.2(h), coral growth follows a slightly irregular firmground of oncolitic packstone.

Underbeds   Underbeds form by the underlying lithology being cemented to the overlying limestone, thus obscuring the sole (and associated trace fossils and sedimentary structures), and generally making it difficult to extract fossils from the underlying sand (right) or mud (below centre). Overbeds may similarly be found at beds tops.

Growth bedding   Growth bedding has not been recognized in ancient sediments. Today it is forming where *in situ* growth is taking place within a unit of free-living calcareous algae. *Lithothamnion* is depicted in Fig. 2.2(l).

Pseudobedding   Pseudobedding is produced by pressure dissolution, stylolite-controlled layering, in generally homogeneous limestones, typically pack-stones to grainstones (Simpson, 1985). There are three types:

- Unverifiable pseudobedding is where a stylolite parts identical lithologies.
- Exaggerated true bedding is where a stylolite separates different lithologies.

- Verifiable pseudobedding is where true bedding (x) can be identified (often cryptic) between stylolite surfaces. At outcrop, stylolites often weather back as prominent surfaces, or splay into anastomizing feathery lines, a situation typical of fine-grained carbonates such as chalk (Fig. 2.2(r)).

Cross-lamination   Intervals of cross-laminated sands with mud flasers (wisps) transitional to wavy bedding to lenticular bedding (lenses of sand). Trace fossils are often distinct on soles but shelly biota is generally sparse. Fig. 2.2(n), centre diagram, shows parallel, millimetre-thick beds, typically of silt grade. Undertracks may be present. Muddy laminae may be associated with shell stringers and pavements. Herring-bone cross-stratification is a good indicator of tidal environments but check whether opposing foreset dips are only apparent. Couplets (typically millimetre-thick sands separated by a thin mud) are also indicative of a tidal regime (lowest diagram in Fig. 2.2(n)).

Ripple-drift lamination   Ripple-drift lamination is generally poorly fossiliferous, even for microfossils; this is due to the poor development of ripple troughs.

Cluster of shells   A 'cluster' of shells may be diagenetically preserved within an early formed concretion; there may be subsequent loss of shells in surrounding sediment.

Cross-stratification   With cross-stratification in tidal sand-grade sediments, trace fossils occur at the base of a bed and rise up foresets corresponding with phases of low sedimentation (often muddy laminae). Shelly fossils, bones and wood are mostly concentrated along bottom sets.

False indications of bedding   False indications of bedding, particularly in cores, may be due to laminar trace fossils such as *Zoophycos*, or even bedding-parallel simple burrows.

Unconformity   (Fig. 7.3)

Biogenic stratification   Biogenic stratification (Meldahl, 1987) and biogenic graded bedding (Trewin and Welsh, 1976) have only been recognized in modern sediments. Both result from the activity of infaunal organisms. In biogenic stratification, fines are eliminated by the activity of polychaetes and arthropods. The lower limit of their activity is relatively sharp but never as sharp as with hydraulic processes. Biogenic graded bedding results from size sorting of grains, especially shells, with the concentration of larger particles in the lower region of the animal's activity.

Concretions and diagenetic structures   Useful indications of stratification are layers of concretions and diagenetic structures such as burrow-filled flints and cherts. The latter relate to an original sedimentary surface well above the actual layer. Diagenetic bedding can be particularly difficult to prove. The origin of thin stratiform concretions is generally clear and a lateral termination to the layer as it passes into normal sediment demonstrates its origin. Commonly, diagenesis has followed primary bedding.

Cyclic sedimentation   Cyclic sedimentation (rhythms when asymmetric) with regular repetition of facies or sequences of facies ranging from millimetres to many metres in thickness can be due to a variety of causes. It is important to distinguish between autocyclic causes, as with lateral shift in facies due to river meandering, from the popular allocyclic orbital controls with periodicities of 19, 23, 41 Ka upwards. Allocyclic controls have been widely invoked to explain apparent climatic cyclicity in the rock record. In the field pay particular attention to the manner of vertical facies change. Evidence for a decelerating flow should be apparent from changes in grain size and bedform (Box 8.1). Increase in wave activity may be due to minor topographical change and more open (marine) exposure, or to shallowing (perhaps one's first thought).

2.4   Graphic logs

A graphic log (Fig. 1.3) provides the best appreciation of a section. It enables evaluation of the facies, and a 'ladder' on which to position the fossils observed and collected. A good log acts as a summary diagram of all the information about a section. It must be readable. For palaeontological purposes, realistic and annotated logs are the most useful. Stylized logs are constrained by the symbols available and can give a misleading impression. For example, if the top of a bed grades imperceptibly from sand to mud, then no filter-feeding organism could have colonized the fine-grained substrate. Drawing a realistic log makes one think of the implications of every line. Software is now available to do this. More detailed logging sheets are used industrially. These cover the sonic log and the γ-log, as well as the grain size, bioturbation index

(BI), etc., all of which are too elaborate to show at textbook size. They can be readily obtained.

Decide whether it is necessary to log the whole sequence, and all at the same detail, or whether a representative log is sufficient. A representative log is more difficult to produce because it assumes one knows which facies are to be selected. Decide whether the log is to be essentially one-dimensional (i.e. core, stream section), or two-dimensional, where an extensive face exhibits lateral variation. The log in Fig. 1.3 was generated for a field excursion. Other examples are given in Barnes (1995) and Graham (1988).

Logging strategy at outcrop

Inspect section   Clean up loose material, remove vegetation to expose important boundaries. Mark important features. It may be necessary to make a grid using string, (bean) netting, chalk or pegs if it is intended to record lateral changes.

Scale   On a log sheet or in a notebook it is not possible to record a unit in less than 2.0 mm. Thus, if the smallest unit to be recorded is 50 mm then the scale required will be 1 mm on the log = 25 mm at the outcrop or core. Details can always be shown at a larger scale. The logging scale (LS) can be obtained from the following formula:

$$1\,\text{mm on log} = \frac{\text{Thickness of smallest feature to be measured (mm)}}{2.0}$$

If the smallest feature is 50 mm thick then

$$1\,\text{mm on log} = \frac{50}{2.0} = 25\,\text{mm}$$

Prepare notebook   Prepare notebook or logging sheets with columns headed according to the information required. Keep a column for notes. If notes are to be recorded on disk or tape, generate a checklist for the description. Test this out if you intend to use computer software to generate the log.

Recording   If possible, work upwards (stratigraphically) rather than downwards. This way it is easier to gain an appreciation of the nature of bed and facies contacts. At the outcrop, first mark in the weathering profile. In siliciclastic sediments this generally corresponds to lithology and grain size, but in carbonate sediment dissolution seams may be present and autochthonous (reefal) units may weather back more than a micrite. Lithology can then be added against the weathering profile or as a separate column, and grain size as a further column. In well-sorted sands, grain size estimation does not present any problem, but in poorly sorted sediments, estimate the mode and then add a point or points to indicate range of grain size present. In carbonate sediments, more emphasis is placed on sorting. Plot wackestone, packstone and grainstone. Add bedding and other sedimentary structures. Lightly colouring a log aids visual impression (e.g. clay in light blue, silt in grey, sand in yellow, gravel in brown). Distribution of fossils on logs presents problems. Rows of symbols are quite unsatisfactory, though some may be added to indicate preservation mode and completeness. Frequency (Section 5.3.8) may be indicated by arranging the columns with frequency decreasing left to right or by line thickness.

Photograph   Photograph the logged section and mark the photographed areas on the log using photo numbers.

Sampling points   Indicate sampling points on the log using sample numbers.

Go over log   Check entries and make a preliminary analysis of the facies; note the presence of cycles and sequences.

2.5 Sampling: two important questions

At what interval should samples be taken?

Attempt to determine the relevant biological processes and their rate of operation. Biological processes that are of most concern are growth, evolution (for biostratigraphy) and ecological change. Many biological processes are too rapid or are unrecordable, e.g. angiosperm fertilization. Here are some typical timescales:

- Speciation may operate over 10^5 to 10^6 years (for biostratigraphic fossils).
- Ecological succession might involve 10^0 to 10^3 years.
- Habitat destruction may be instantaneous to 10^3 years, and regional extinction takes place in 10^3 to 10^4 years.

What size of sample is needed?

The sample size needs to be chosen so it will yield data that identify the pertinent patterns of biological or taphonomic processes and attributes (community identification, biostratigraphy, morphological variation). Although not essential for the field, it is often useful to be aware of the statistical tests that can be applied to the samples, especially if the number of samples is less than 30 (Appendix B).

Sedimentation rates for various environments are shown in Fig. 2.3 with a range of values for each to

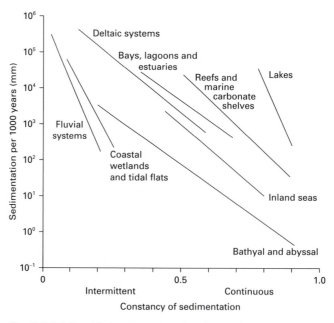

Fig. 2.3 Relationship between rate of sedimentation (millimetres of sediment per 1000 years) in various modern sedimentary environments, and constancy of sedimentation, i.e. fraction of time during which sediment actively accumulates. Adapted from Schindel (1980).

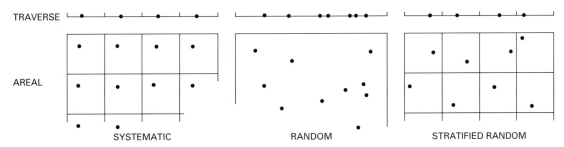

Fig. 2.4 Traverse and areal sampling schemes.

Table 2.1 A table of 100 random numbers.[a]

72	20	09	65	77	70	94	34	25	85	96	11	78	07	30	61	03	04	63	92
06	01	76	27	90	68	71	38	29	23	13	59	93	52	83	44	53	37	91	21
57	79	86	47	80	35	81	05	62	50	33	15	73	39	42	60	55	19	56	87
48	40	64	67	16	12	98	32	84	69	22	17	24	28	02	97	66	45	41	43
58	00	54	46	82	74	36	49	31	14	89	08	18	95	88	99	10	75	51	26

[a] Use single numbers as spacings (e.g. cm, mm) or pairs of numbers as coordinates.

allow for aggradation and erosion. But use the figures with caution. Event beds normally need be sampled only 'once' regardless of thickness. But ascertain just how the biota is distributed through an event bed.

Attempt to determine the frequency of a taxon per unit volume or per unit weight of sediment, or determine the frequency of the taphonomic effect, e.g. valve dissociation. For bulk samples of macrofossils it is best to determine a sampling curve (Appendix B.6) corresponding to size of sample and number of species collected. For instance, in the Silurian of Wales, increase in species number fell off at a total of 200 individuals, so any sample should contain at least 200 individuals. Search along the outcrop for rare species that may be important for reconstructing higher trophic levels. In many instances one will be searching purposefully for particular modes of preservation, or for stratigraphically useful fossils. No statistical inferences can be made from such collections about distributions or trends. The axiom of statistical sampling is that it should be free of bias. At outcrop, consider these questions:

- What is to be determined?
- Why is it not feasible to measure everything?
- What degree of precision is required?
- To what extent will the sampling pattern be controlled by lack (gaps) of exposure, accessibility, differential weathering?
- What sampling method is required?
- What is the budget? Sampling on a regular grid is cheapest (Fig. 2.4). Duplicate samples can be taken.

- What are the relative sampling and analytical costs? And what are the relative errors involved? Errors in sampling can be quite high.

For *trends* over an area or along a transect, e.g. decrease in size, decrease in frequency of a taxon or changes in diversity, adopt a *systematic* or *stratified random* sampling scheme (Fig. 2.4). Systematic sampling is useful when sampling points are to be spread evenly, and it is a simple method. If systematically sampling, select the first sample position randomly. For distributions, e.g. orientation or the amount of cover of different taxa on a bedding surface, use a simple *random* sampling scheme. If it is intended to sample a particular facies, perhaps only the limestones of an alternating limestone – shale succession, then take samples at regular or random intervals, as the case may be, above the base of successive limestones. When sampling to test similarity or difference, adopt the same sampling scheme for each unit; one sampling scheme might be to decide whether the taphonomic make-up of two beds is similar. To generate random intervals, a set of random numbers (Table 2.1) can be used, or a pair of dice may be thrown.

If a quadrat is used then its size should be related to the size of the fossils (plant fragments or shells). Size can be determined with a pilot study, by taking quadrats of increasing magnitude and plotting the increase in number of taxa recorded. Quadrats may be marked out with chalk, or staked out with skewers or tent pegs. A quadrat with a 0.5 m side will be suitable for most situations. The quadrat frame is covered

Table 2.2 Distribution of microfossils in sediments.[a]

	Ostracodes	Chitinozoans	Radiolarians	Calpionellids	Conodonts	Foraminifera	Dinocysts and acritarchs[b]	Coccoliths	Spores and pollen[b]	Diatoms
Mudrocks (dark)[c]	+	+	−	−	−	+	+	−	+	−
Limestones[d]	+	−	−	+	+	+	+	+	−	\
Sandstones	−	−	\	\	−	−	−	\	−	\
Coal, lignite[e]	\	\	\	\	\	\	\	\	+	\
Siliceous rocks[f]	−	\	+	\	−	−	−	−	−	+

[a] Key to symbols: (+) abundant, (−) rare, (\) generally absent.
[b] Organic-walled microfossils are destroyed by weathering, which can penetrate deeply.
[c] Mudrock includes claystones, mudstones and clayey siltstones ± calcareous varieties.
[d] Including chalks; fossils much reduced in dolomites.
[e] Increasing thermal grade leads to decrease in recovery.
[f] Of primary nature: chert, flint, diatomite.
Source: Bignot (1985).

with a clear plastic sheet on which 100 randomly distributed points are marked. Record the identity of each fossil at each point. The data can be analysed on a percentage basis or by ranking (e.g. Domin scale where + = a single individual, 1 = 1–2 individuals; 2 = less than 1%; 3 = 1–4%; 4 = 5–10%; 5 = 11–25%; 6 = 26–33%; 7 = 34–50%; 8 = 51–75%; 9 = 76–90%; 10 = 91–100%).

2.5.1   Microfossils and palynofacies

For microfossil and palynofacies analyses, only material that is *in situ* should be used for stratigraphic purposes. It is vital that the sample is unweathered and clean and, for pollen and spores, has not been exposed to the air. Use new polythene sample bags, or tubes, with a closure. The distribution of microfossil groups in different lithologies is shown in Table 2.2. Lithologies are given in general terms. For instance, foraminifera will only be found in marine sandstones, unless derived. Metamorphosed sediments are omitted but may yield rare instances. Lenses of limestone or chert between lava pillows are important sources. Take spot samples at regular intervals, or over a regular number of events (allow for amalgamation of turbidites; Fig. 6.2), or 'channel' samples, where sampling is continuous over a metre or two (thus reducing the possibility of accidentally sampling unfossiliferous intervals). Sample size will depend on the nature of the sediment. As a general guide, for ostracodes and foraminifera in fossiliferous mudstones, etc., 200–500 g will yield sufficient specimens. For coccoliths 30 g is sufficient. But, for sampling highly 'diluted' sediments, e.g. conodonts in an encrinite, several kilograms may be needed for later acid digestion. In bioturbated sediment (including most chalks), there may be no point in sampling at closer intervals than 0.5 m for stratigraphic purposes because of the redistribution of the biota by the burrowing activity. If possible, sample unbioturbated sediment. Samples for later palynological analysis (palynofacies) may be important for correlation and environmental interpretation, and they are especially important for environmental interpretation when macrofossils and sedimentary structures do not give a clear indication (particularly in cores). Fig. 2.5 shows the potential of palynofacies analysis in coastal facies. Marine dinocysts are reported to have been blown from beaches far inland today. This may also have happened in the past! When sampling for foraminifera and ostracodes, consider how the lithology may have been affected by size sorting prior to burial (Box 6.1).

2.5.2   Condensed deposits

Condensed deposits require particularly careful sampling for stratigraphic purposes. With condensed mudrocks, mark and map the rock face. Note carefully from where each sample is extracted, and number each sample immediately. Often there will be subtle changes in lithology, associated with differences in mode of preservation. In condensed limestones (e.g. Ammonitico Rosso, Cephalopodenkalk, many hardgrounds), adopt the same procedure and collect large and overlapping samples for later subsampling (Table 7.4).

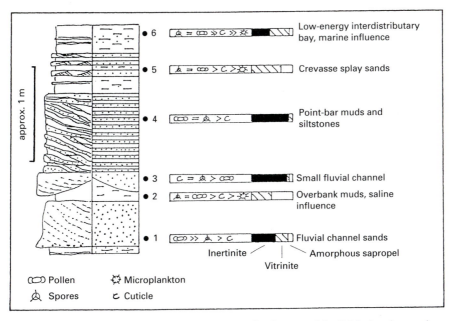

Fig. 2.5 Simplified palynofacies analysis of deltaic distributary – interdistributary bay environments. Basal channel sands (1, 3) with dominant inertinite (e.g. charcoal). Muddy sediments above (2) contain some marine microplankton. Point-bar facies (4) contain inertinite with cuticle and vitrinite increasing upwards. In (5, 6) marine microplankton indicate lagoonal rather than lacustrine environments and closer analysis indicates a degree of salinity. On sedimentary structures alone the marine influence would possibly be unrecognized. The figure is based on the Jurassic of Yorkshire, England; other palynological classifications are possible. Oil yield is proportional to amount of sapropel. Adapted from Fisher and Hancock (1985).

2.5.3 Fossil plants

Sampling techniques for fossil plants pose more problems than for autochthonous animal remains since, apart from horizons with just roots, it is uncommon to find autochthonous material that combines roots and stems. Use can be made of drifted plant accumulations (generally good in type 10 deposits) if the limitations of such material are appreciated. There are several reasons why the results differ from those obtained by plant ecologists:

- The cover does not represent a living community.
- The cover is unlikely to be a single bedding surface: rocks rarely split perfectly.
- The quadrat can rarely be placed in a random manner.
- The way the rock splits is influenced by the type of plant present.

In the last item, note how the 'cover' for each taxon relates partly to the original biomass and partly to the preservation potential and hydraulic properties of the original plants. If analyses made for sections of similar facies show consistent plant assemblages, then the assemblage is a positive feature of that facies, and can be used in environmental interpretation.

2.6 Dominantly autochthonous buildups

Autochthonous buildups include reefs, patch reefs, banks, bioherms and biostromes (Tables 2.3 and 5.2). The principal aims are to determine the overall morphology, detailed anatomy and evolutionary history of the buildup.

2.6.1 Potential problems

Species diversity Organisms responsible for buildups have shown a remarkable diversity and undergone much change over geological time (Fig. 2.6). Moreover, the ecology of most groups has changed over geological time. For instance, Palaeozoic rugose corals were heterotrophs and often only a minor part

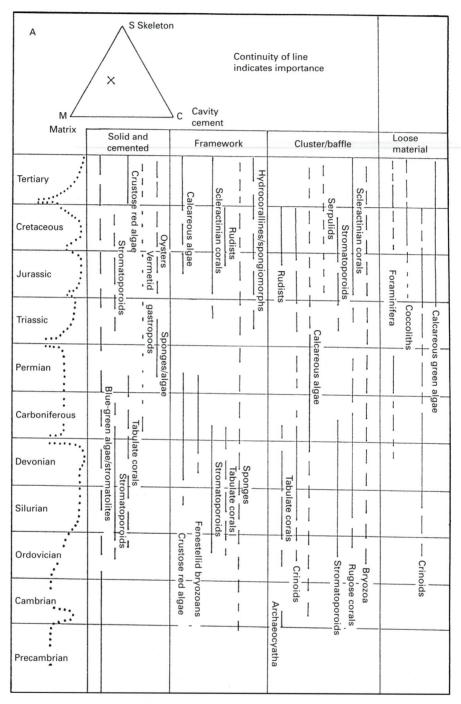

Fig. 2.6 (A) Distribution of organisms through time that form (1) autochthonous and parautochthonous buildups (an attempt is made to divide the groups on their role as forming (a) solid and cemented buildups, (b) frameworks, (c) clusters and baffles); (2) autochthonous and allochthonous accumulations of loose skeletal material (including coccolithic chalk). The dotted line (left) gives an indication of the importance of autochthonous buildups through time, with periods of their apparent absence. In the triangular plot for components of autochthonous buildups, the cross marks the buildup depicted in Fig. 1.3.

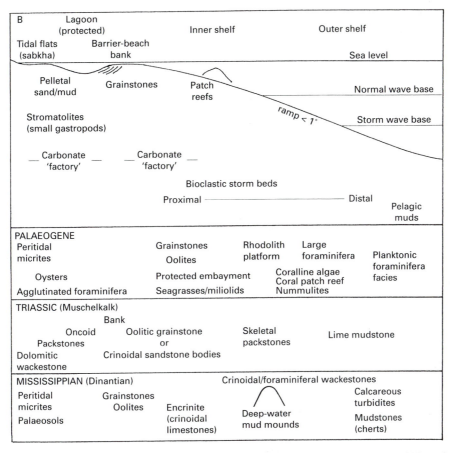

Fig. 2.6 (B) Generalized cross-section across a carbonate platform and distribution of biotas/facies for the Palaeogene (after Buxton and Pedley, 1989), Triassic (after Aigner, 1985) and Dinantian (Mississippian) (after Simpson, 1987). According to Read (1985), modifications to the platform may include (1) distal steepening; (2) reef belt; (3) outer (rimmed) margin.

of the biota in low to middle latitudes, whereas many younger scleractinian corals are photosymbionts restricted to oligotrophic environments and flourishing in low latitudes. It is useful to distinguish two groups: oligotrophic photosymbionts, or *Chlorozoans*, and heterotrophs, or *foraminifera–mollusc* associations.

Unstable mineralogy Some organisms (e.g. scleractinian corals) have skeletons composed of relatively unstable mineralogy (aragonite), which readily undergoes dissolution or replacement (Appendix C), and in others skeletal elements are only loosely bound together during life and dissociate quickly on death, and are then easily transported. Bedding and anchoring devices (crinoid roots) may provide clues to the mode of accumulation.

Complex stratigraphy There is the problem of unravelling the stratigraphy of a complex buildup and determining time lines. It is important to know what the relief of the 'reef' surface was like. Search for truncation surfaces and local blanketing. Relief may be due to a pre-existing structure, or have been enhanced by differential compaction. Penecontemporaneous erosion may have taken place during formation (check for truncation, evidence of emergence).

Facies differentiation Any but the smallest buildup displays facies differentiation, e.g. core and flank facies. Within these facies there will be further differentiation, e.g. intraframework sediment, local patches of skeletal sand, local cavities.

Table 2.3 A classification of buildups (see also limestone classifications).

	STRUCTURE				
	Matrix	Allochthonous material (%)	Diversity	Talus	Cavities (%)
Cement reef skeletons cemented	low	low	generally low	may be present	generally low
Frame reef skeletons with local contact	variable	can be high	generally high	present	can be high
Cluster reef skeletons are generally discrete	high	can be high	generally high	common	can be high
Stromatolitic/ thrombolitic	micrite	low	low	channelized	variable
Mictritic mud mound	–	low	variable	absent	variable
Loose material crinoidal fusulines nummulites[a] branching rhodoliths	high/low	variable	low	absent	can be high but even size

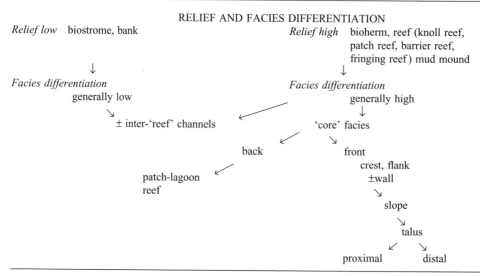

RELIEF AND FACIES DIFFERENTIATION

Relief low   biostrome, bank

Relief high   bioherm, reef (knoll reef, patch reef, barrier reef, fringing reef) mud mound

Facies differentiation
generally low

± inter-'reef' channels

Facies differentiation
generally high

'core' facies

back   front

patch-lagoon reef

crest, flank
±wall

slope

talus

proximal   distal

[a] Mainly allochthonous.

Cementation   In well-cemented buildups, particularly micritic buildups, it will not be possible to resolve the facies or the composition in detail in the field. Sample for slabbing and preparation of peels and thin sections, using a regular pattern in order to detect facies differentiation.

Skeletal attack   As many animals grow, their skeletons come under attack from a range of organisms, from algae to clams, that penetrate exposed skeleton (e.g. undersurface of corals, or dead areas), thereby tending to destroy the buildup. Borers are generally obvious in substantial skeletons, but burrowers can almost completely disrupt more fragile skeletons, such as laminar rhodophytes (Fig. 2.7).

Dissolution or dolomitization   Skeletal buildups are rather porous. They often formed close to sea

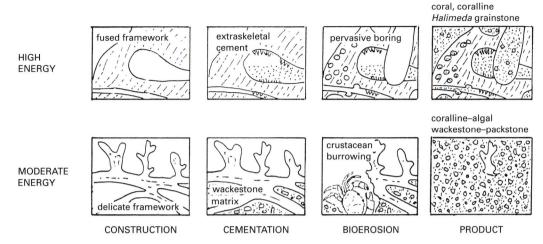

Fig. 2.7 Taphonomy of a modern coralline-algae buildup in high and moderate energy environments. Bosence was able to observe the changes that took place from the living alga to final burial. In a fossil example the original life form has to be deduced from the observed product! Adapted from Bosence (1985).

level. Fluctuations of sea level may have led to flushing by seawater or freshwater leading to dissolution or dolomitization. Extensive dolomitization is a common feature of reef cores and may make work difficult.

Large buildups With larger buildups one is involved with a substantial three-dimensional structure. The exposures (quarry, coastal cliff, etc.) may be difficult to work and the degree of weathering insufficient, or joint facies may be spar-covered.

Differential growth Differential growth of the buildup relative to the adjacent areas probably affected turbulence, food availability, light, etc. Anticipate that there may be upward and lateral changes in the composition of the biota, and in growth forms (Figs 5.9 and 5.11).

Sea-level changes In general, the maxim that *organic growth can keep up with subsidence* holds good (e.g. Fig. 1.2). Thus one can anticipate upward shallowing, increasing wave activity and possible changes of growth form. Any relative change in sea level that took place will have significant effects: flooding, hardground formation, karst, etc.

Linear buildups If the facies either side of a linear buildup differ, this will have a bearing on any interpretation of the depositional environment. Many

smaller buildups in the Palaeozoic and Mesozoic do not appear to have acted as a physical barrier.

Reefs Wood (1995) has emphasized the manner in which buildups and patches are formed. Aclonal and solitary organisms form quite small buildups with low relief. Clonal/modular scleractinian corals and/or encrusting calcified algae often form buildups, which have large areas and large topography; colloquially known as *reefs*, they are forming today in low-nutrient (oligotrophic) settings. In a reef, the sheltering cryptic community tends to have a high diversity.

2.6.2 Field strategy

1 Make a reconnaissance.
2 Map an outline of the buildup at a suitable scale (1 : 10 or down to 1 : 1).
3 Map the major facies, e.g. core (Fig. 2.8) and flank facies. Tracing bedding surfaces in the flank facies will indicate the topography of the buildup.
4 Establish that there is a major degree of autochthony (Section 5.2.1).
5 Establish major time lines and event markers, major discontinuities, ash bands, storm layers and emersion events. These provide a basis for unravelling the evolution of the buildup. Note that autogenic ecological succession cannot generally take place across such events (Section 5.2.2).

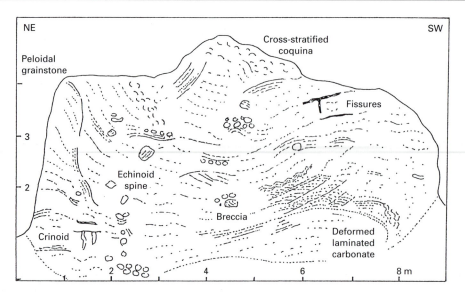

Fig. 2.8 Field sketch of core facies for a mud-mound buildup in the Dinantian (Lower Carboniferous) of Derbyshire, England. It was not possible to establish time lines over the area. Adapted from Bridges and Chapman (1988).

6 Analyse the principal skeletal components:

(a) Map selected areas using a quadrat method (Section 2.5).

(b) Attempt to identify the principal skeletal components, their role (Table 2.3) and mineralogy (Appendix C). Unfamiliar types of fossil may be present, or unfamiliar modes of preservation. It is important to appreciate the implications of mis-identification. For example, a simple alga might have tolerated a wide range of salinities and emergence, but a sponge or coelenterate would have had a much narrower ecological tolerance.

(c) Identify the growth forms and their ecological significance (Table 2.3), determine the size and cover of individual components using a gridded overlay. The results may be used to determine percentage vertical and lateral change. Identify the type of substrate below each component, whether formerly soft, or whether one component is encrusting another, or a clast.

(d) Recognize stages of development (Fig. 5.9), or significant sequential changes in taxonomic composition and growth form.

(e) Search for growth banding (Fig. 1.2) to establish growth rate and productivity (which may vary vertically and laterally), and assess the time represented by the buildup.

(f) Attempt to assess and tabulate the relative resistance to breakdown of the components (Table 3.1), and the role of skeletal destroyers (bioeroders).

7 Interframework biota (matrix) may be composed of several size fractions, and include organisms that, mobile or fixed, nestled within the framework, together with skeletal elements from organisms that readily dissociated on death (e.g. brittlestars), material washed in (biogenic and terrigenous), and faecal pellets, etc. Analysis will allow one to 'complete' the picture for a reconstruction. If the matrix is softer than the framework (as is often the case at Palaeozoic outcrops) then skeletal components may weather out. There may be a rich microbiota to sample. If the matrix is harder than the framework, carefully examine the rock and collect for later slabbing and sectioning.

8 The nature of cavities is important ecologically and sedimentologically (Fig. 2.9). Cavities are now (a) still open or oil-filled, (b) spar-filled, (c) sediment-filled, (d) filled with crystal silt, or a combination of these. Try to determine the three-dimensionality of cavities. (This may require sampling and slabbing.) First, investigate the nature of any open cavity. Then tackle filled cavities. Note geopetal structures giving indication of depositional slope. Primary cavities result from (1) natural cavities (cephalopod camerae,

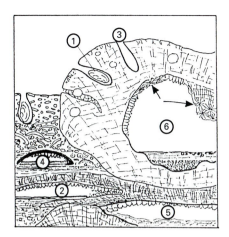

Fig. 2.9 Cavities in buildups: (1) primary cavity of bivalve shell and intraskeletal spaces, (2) irregular growth cavity, (3) borings, (4) umbrellar structure, (5) compactional cavity, (6) diagenetic cavity.

bivalve shell), (2) irregular growth (mm scale for lamellar calcareous algae to cm–dm scale with laminar corals or stromatoporoids), (3) animal or algal borings, (4) umbrellar (shelter) structures. Compactional cavities (5) are common below tabular skeletons and usually display geopetal sediment. Diagenetic cavities (6) occur especially where a skeleton has been wholly or partly dissolved. This may not be easy to prove in the field. Try to determine what was originally present. Look at the margins (arrowed, Fig. 2.9), where impressions of skeletal features may be preserved. A clue to dissolution and replacement having occurred is an excessive proportion of spar. For instance, common centimetre-size botryoidal drapes may represent dissolution of algal rhizomes below thin curved laminar crusts. Be prepared for larger-scale dissolution features associated with emersion.

9 There is only a limited amount that can be done in the field to appreciate diagenetic history. Try to detertmine the order in which major dissolution and cementation events occurred. For example, if moulds of framework skeletons are undeformed, then matrix cementation took place before skeletal dissolution.

2.7 Bedded fossiliferous sediments

The first objective, following mapping and logging, is to establish whether the units contain autochthonous,

parautochthonous or allochthonous fossils (Section 5.2.1). Most outcrops of lithified sediment display bedding surfaces or sections normal to bedding. The information that is potentially available from each is different but complementary. In soft sediments the sections are nearly always normal to the bedding. Particular attention is then required to obtain the information on bedding (surfaces).

Bedding surfaces (including quarry floor and shore platform) may help with recognition of colonization surfaces and smothered (obrution) assemblages, analysis of fossil articulation, orientation, pockets, lenses, clusters, lateral variation, diversity, analysis of trails and bedding-parallel trace fossils; they may also provide an opportunity to collect larger and flatter fossils.

Sections normal to bedding provide information on the detailed stratigraphy; they provide an opportunity for spot and channel sampling, and to examine tiering of body and trace fossils, shell articulation and orientation and vertical and lateral changes in the biota.

2.7.1 Potential problems

1 In sections normal to bedding, a well-cemented limestone does not display the fauna adequately or its distribution. Sample oriented blocks for slabbing.
2 If only sloping dragline surfaces are available, trench for the stratigraphy and attempt to split blocks for bedding information.
3 Where bedding surfaces are found only as horizontal clefts, to extend the upper and lower faces will require longwall mining – a dangerous operation.
4 Bedding surfaces found only as vertical clefts can be extended with care.
5 Where scree has to be traversed in order to examine an exposure, remember that material may be falling from above. Check that no danger is being created to anyone below when clearing the face. Do not trench in loose sediment without adequate precaution against cave-in – collapse occurs more quickly than one can evacuate.
6 If the 'bedding' surface is a dissolution surface (pseudobedding) then search for true bedding.

2.7.2 Field strategy

1 For allochthonous accumulations, collect bulk samples for later analysis of frequency, diversity and 'time averaging'. For information concerning

the source of the fossils, it may be possible to locate a site where the elements are autochthonous. Also search for information that gives a clue about derivation, such as attached sediment, type of sediment infilling cavities, and attempt to infer the life environment from morphological evidence.

2 For autochthonous and parautochthonous accumulations, seek as much ecological information as possible by systematically checking through (Section 5.2). Record observations on the life attitude and the type of trace fossil (Section 8.4). Collect bulk samples for analysis (size distribution, morphological information, growth rates, trophic structure and biomass). Search along the section for rarer elements that may be important in upper trophic levels. Collecting is easy in facies where shells weather out readily (e.g. muddy limestones). But consider the validity of any such collections for community analysis. With extensive surfaces, quantitative analysis is required when investigating abundance, dispersion, diversity and population structure.

3 Where bedding surfaces are extensive, lateral variation in the biota is to be expected. At hardgrounds look for crevices and pockets that may yield a different but related biota. Attempt to establish, from trace fossils and encrusters, whether or not a hardground displays evidence for gradual lithification and continuous colonization with gradual faunal (autogenic) succession. Most shallow-water marine hardgrounds display faunal replacement, perhaps indicating an emersion event. Be prepared for larger-scale features associated with emersion, e.g. caves and cave deposits.

4 Much information can often be obtained from scree material, or boulders on a beach, but scree and boulders have to be 'fitted' to the stratigraphy.

2.8 Bioturbated sediments: trace fossils

Loose sand or silt with minor amounts of mud/ clay   Never apply water, by spraying or a sponge, to surfaces, especially core; the sediment will disintegrate and the core will become virtually useless. Use a soft brush, knife, sharp-edged trowel or onion hoe to form a smooth face. On a vertical face, work downwards to stop fine sand settling over the lower part of the face. Analysis of the ichnology can then follow, accompanied by sketching and photography. It is useful at this stage to select areas that are suitable

for making peels (Appendix A). The course of small burrows can be determined by scraping the face, but any scraping is a destructive process. It may also be possible to get three-dimensional information from surfaces parallel to the bedding or cross-bedding. Muds tend to smear and do not peel well.

Firm sand and silt and thicker mudstones These common lithologies are difficult to work and they are generally too tight to scrape and dissect satisfactorily. Furthermore, the rock does not readily break, at least not in the way one hopes. Local cementation may make a smooth face difficult to obtain, and the stiff mud-filled burrows or lenses tend to smear. Spraying or natural rainstorm, or gentle washing by a stream may provide differential exposure of trace fossils and sedimentary structures.

Lithified siliciclastics and mudrocks   Trace fossils will be generally exposed on bedding surfaces and faces. The rock may split to reveal partial surfaces with different toponomic expressions, or undertracks (Section 8.2). Take time to evaluate each slab to determine whether a complete trace can be reconstructed. It may be necessary to collect blocks for later cutting and slabbing. Slabbing normal to the burrow length will give most information. Label specimens as they are collected with 'top' or an arrow pointing stratigraphically upwards. Indicate orientation on sketches and the orientation of faces that have been photographed. Much information can generally be found on loose slabs. Turn them over to observe both sides, but use a hammer in snake or scorpion country!

Limestones   Limestones pose particular problems for ichnology because of their diagenetic history (silicification, dolomitization) and early (rapid) lithification, and the general lack of colour contrast to show up traces and ichnofabrics. There is also the problem of dissolution and the formation of stylolites. Micritic burrows or the micrite lining to a burrow (e.g. *Palaeophycus*) may be recrystallized to spar. Heterolithic and bedded grainstones and micrites or argillaceous limestones may be considered in the same way as lithified sands and silts, though underbeds and overbeds (Fig. 2.2(k)) may make life difficult! Hardgrounds are generally readily identified by their encrusters and borings, and are always important stratigraphically. Skeletal elements inhibit bioturbation, but their distribution through a bed gives a useful clue to the degree of bioturbation. Thick (metre-scale) beds of shallow marine limestones are nearly always composite. Decimetre samples will be required for later slabbing to ascertain primary bedding. Sketch and photograph the face before removal and label the samples, especially their orientation. Tiering in calcareous turbidites is often clearer than in siliciclastic ones. Chalks pose particular problems, especially when pure. Spraying the smoothed surface with a solution of methylene blue or applying a light film of thin oil is useful. Allochthonous chalks can be identified by nodules with traces at abnormal attitudes, borings. Shear planes and 'horsetail' dissolution structures can strongly resemble burrows.

2.8.1 General strategy

1 Following a reconnaissance examination of the site, try to determine what primary sedimentary structures are present (Section 8.9). Logging can then be started.
2 The log will include determining the bioturbation index (Table 8.2) and identifying as far as possible the ichnotaxa and their cross-cutting relationships to establish the tiering styles present. It is seldom that all the ichnotaxa will be identifiable, such as finer burrows (*Chondrites* or *Phycosiphon*).
3 Sketch and model the styles of colonization and tiering.
4 Tracing the face laterally may reveal preservational differences and patchiness of the ichnofauna.
5 Sample with respect to the likely size of the ichnotaxa. (This is obviously impracticable with large and deep-tier traces.)

2.9 Core analysis

First, some dos and don'ts. The cost of recovering core from deep wells is extremely high! It is a priviledge to have a sight of such core, and entails a high degree of responsibility in treating the core and in its logging. The state of core can vary very much. Some will be in plastic sleeves or have a cake of drilling mud on the outside. This has to be cleaned off. The core may be stacked in boxes, which are often heavy, or laid out on benches. The lighting will also vary depending on location. It may be necessary to use additional lighting. The core may have been slabbed or sliced and resin mounted. If the core is resin mounted then consider oneself fortunate. It may already have been photographed. Top and bottom of each length of core and core depths may or may not

be clear. There may be driller's labels poked into breaks with the driller's depth, and there may be indications of unrecovered intervals. Driller's depths may not tally with the well depth. The core may be complete or fragmentary. It may have been highly fractured. Top and bottom may be indicated by arrows (generally pointing upwards), or by a pair of coloured lines, such as blue and red, with blue on the left if the top is directed away from the observer. Check these out. Sometimes sections of core can be misplaced in the boxes with the wrong orientation.

- **Do not** remove sections of core from the box, and certainly not without marking attitude and depth, and gaining permission.
- **Never** take samples away without obtaining permission.
- **Never** wet loose sediment or clay, especially swelling clays; they will disintegrate.

Slabbed, lithified core may be lightly moistened with water to display the lithology better. Use a spray, sponge or paintbrush of suitable width. Other techniques are available.

If only a narrow core is available, there may be features that are puzzling, especially from deep wells with extensive fracturing. It is useful to take this into consideration so as to anticipate how depth of burial and thermal history have affected the mineralogy and diagenesis. The core may look very different from more familiar rocks of similar age at outcrop that have experienced different burial histories. In particular, deformation and fracturing may have taken place, and perhaps also shell dissolution.

The strategy in core analysis is essentially the same as for outcrop analysis. First check how the core is to be laid out or has been laid out. Are any boxes misplaced? Are boxes with pairs of cores in their correct position? Make a reconnaissance by walking out the core, noting where facies changes or contacts occur. It may be useful to flag them with slips of paper.

Before starting to log, make sure the logging sheet is suitable for the facies present. Core photography is tricky. Slabbed core generally needs moistening. Water dries quickly and is absorbed. Light oil or a light coating of a hand cream are useful, and it will be necessary to illuminate the core with lamps (or flash) at 45° to avoid reflection. Some core laboratories have sophisticated camera set-ups. Always place a scale alongside the interval photographed, accompanied by a record of the attitude and the depth.

Caving can be a problem and much depends on the driller's skill.

References

Aigner, T. (1985) *Storm depositional systems: dynamic stratigraphy in modern and ancient shallow-marine sequences.* Springer, New York

Barnes, J. W. (1995) *Basic geological mapping*, 3rd edn. J. Wiley, Chichester

Bignot, G. (1985) *Elements of micropaleontology.* Graham & Trotman, London (Contains advice on sampling and a good review of microfossils)

Bosence, D. W. J. (1985) Preservation of coralline–algal reef frameworks. *Proceedings of the Fifth International Coral Reef Congress, Tahiti, 1985* **6**: 623–28

Bridges, P. H. and Chapman, A. J. (1988) The anatomy of a deep water mud-mound complex to the southwest of the Dinantian platform in Derbyshire, UK. *Sedimentology* **35**: 139–62

Buxton, M. W. N. and Pedley, H. M. (1989) A standardized model for Tethyan Tertiary carbonate ramps. *Journal of the Geological Society* **146**: 746–48

Fisher, M. J. and Hancock, N. J. (1985) The Scalby Formation (Middle Jurassic Ravenscar Group) of Yorkshire; reassessment of age and depositional environment. *Proceedings of the Yorkshire Geological Society* **45**: 293–98

Graham, J. (1988) Collection and analysis of field data. In Tucker, M. E. (ed.) *Techniques in sedimentology.* Blackwell, Oxford, pp. 5–62

Meldahl, K. H. (1987) Sedimentologic and taphonomic implications of biogenic stratification. *Palaios* **2**: 350–58

Read, J. F. (1985) Carbonate platform facies models. *Bulletin of the American Association of Petroleum Geologists* **69**: 1–21

Schindel, D. E. (1980) Microstratigraphic sampling and the limits of paleontological resolution. *Paleobiology* **6**: 408–26

Scott, A. C. (1979) The ecology of Coal Measure floras from northern Britain. *Proceedings of the Geologists' Association* **90**: 97–116

Simpson, J. (1985) Stylolite-controlled layering in an homogeneous limestone: pseudo-bedding produced by burial diagenesis. *Sedimentology* **32**: 495–505

Simpson, J. (1987) Mud-dominated storm deposits from a Carboniferous ramp. *Geological Journal* **22**: 191–205

Trewin, N. and Welsh, W. (1976) Formation and composition of a graded estuarine shell bed. *Palaeogeography, Palaeoclimatology, Palaeoecology* **12**: 219–30

Wood, R. (1995) The changing biology of reef-building. *Palaios* **10**: 517–29

Further reading

Field mapping and surveying

Barnes, J. W. (1995) *Basic geological mapping*, 3rd edn. J. Wiley, Chichester

Butler, B. C. M. and Bell, J. D. (1988) *Interpretation of geological maps.* Longman, Harlow (Includes a summary of the rate of present-day movements of the Earth's surface)

Compton, R. R. (1985) *Geology in the field.* J. Wiley, Chichester (Much information on mapping techniques, but thin on palaeontology)

Moseley, F. (1981) *Methods in field geology.* W. H. Freeman, New York

Sedimentary environments and facies

Collinson, J. D. and Thompson, D. B. (1989) *Sedimentary structures*, 2nd edn. George Allen & Unwin, London

Einsele, G., Ricken, W. and Seilacher, A. (eds) (1991) *Cycles and events in stratigraphy.* Springer, New York

Fuchtbauer, H. (ed.) (1988) *Sedimente und Sedimentgesteine*, 4th edn. Schweizerbart'sche, Stuttgart

Lewis, D. W. and McConchie, D. (1994) *Practical sedimentology*, 2nd edn. Chapman & Hall, New York

Lewis, D. W. and McConchie, D. (1994) *Analytical sedimentology.* Chapman & Hall, New York

Reading, H. G. (ed.) (1996) *Sedimentary environments: processes, facies and stratigraphy*, 3rd edn. Blackwell Science, Oxford

Reineck, H.-E. and Singh, I. B. (1980) *Depositional sedimentary environments.* Springer, New York

Scholle, P. A., Benout, D. G. and Moore, C. H. (eds) (1983) *Carbonate depositional environments.* Memoir 33, American Association of Petroleum Geologists, Tulsa OK

Selley, R. C. (1988) *Applied sedimentology.* Academic Press, London

Tucker, M. E. (1991) *Sedimentary petrology: an introduction*, 2nd edn. Blackwell, Oxford

Tucker, M. E. and Wright, V. P. (1990) *Carbonate sedimentology.* Blackwell, Oxford

Walker, R. G. and James, N. P. (eds) (1992) *Facies models: response to sea level change.* Geological Association of Canada, St John's NF

Zimmerle, W. (1995) *Petroleum sedimentology.* Kluwer Academic, Dordrecht

CHAPTER 3

Taphonomy of body fossils

It is the understanding and appreciation of taphonomy that distinguishes a palaeontologist from a biologist.

This chapter attempts to explain the processes and
products of animal and plant fossilization. We are
still learning about the processes of soft tissue preservation, and the exciting new discoveries that are
turning up regularly show that the understanding
of all taphonomic processes is fundamental to
palaeobiology. Skeletal composition is tabulated in
Appendix C, and trace fossil taphonomy is considered
in Chapter 8.

3.1 Introduction

Any body fossil that might be found has undergone
a number of overlapping but consecutive processes
(taphonomy) since it was a living organism (Fig. 3.1).
To begin with, there is death (it is often clear how this
took place, e.g. by burial or predation) and generally
the decomposition of soft tissues, and the possible
break-up of parts. Then follows the sedimentational
history – sorting, damage, reworking – and stratinomy
of the material, leading to its final incorporation into
the sediment, as well as the chemical changes and
compaction that take place mainly within the sediment (diagenesis). Tectonic and weathering processes
may also occur before the fossil is collected. All
these processes tend to modify the original material
(Figs 3.2 and 3.3).

All organisms are made up of a variety of parts
and a variety of materials. The materials (cytoplasm,
sheath, aragonite or calcite skeleton, chitin, lignin,
etc.) vary in their ability to withstand degradation,
dissolution, breakage or chemical change. The parts
(e.g. integument, pygidium, leaf, coxa, dactylus, plate,

ossicle) also vary in their ability to withstand dissociation, compaction and entrainment, and transport
by hydraulic or wind action.

The eventual mode of preservation also depends
on the nature of the sediment in which the organism
is entombed; its ability to take impression detail (compare mud to coarse sand) and its permeability and
cementation potential. Also important are the preservation potential of the sediment, and the basinal and
tectonic histories of the site, particularly those relating
to deformation and to the temperatures that are attained
on burial, e.g. the thermal alteration of conodonts,
graptolites and plant materials (Stach, 1982).

Thus, there are the intrinsic factors relating to the
organism, and the extrinsic factors relating to the sedimentary environment and to its geological history.
Since sediment accumulating in a specific sedimentary environment is likely to undergo a predictable
burial history, any associated biota should have a predictable taphonomy (Box 3.1). Fossiliferous sediments
that have similar taphonomic histories are known as
taphofacies (Brett and Baird, 1986; Speyer and Brett
1988). Features to identify include (a) the likelihood
that the sediment may be reworked, leading to dissociation and fragmentation (but see Table 3.1), (b)
the potential of the sediment to infill cavities, (c)
the potential for phosphate and pyrite to form, (d) the
compaction potential, and (e) the dissolution and
replacement potentials of the minerals involved. The
fossils may locally interfere with, and modify, this
sequence and gradient because of their large size,
porosity (wood, bone), strength and chemistry, etc. In
general, the taphonomic histories of shell beds have
been independent of evolutionary effects such as shell
breakage due to crab nipping (Fig. 5.8).

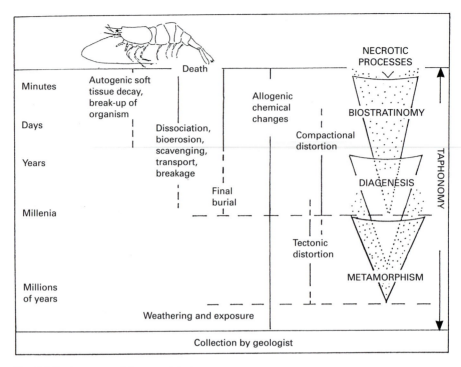

Fig. 3.1 Taphonomy and its associated processes.

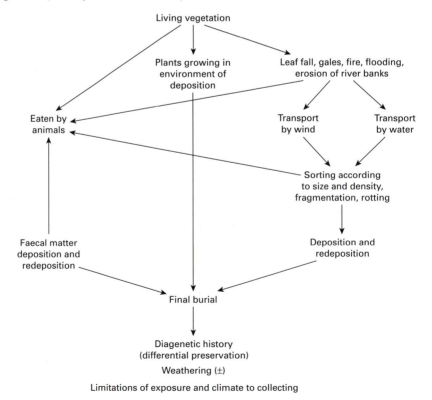

Fig. 3.2 Simplified outline of vegetation taphonomy.

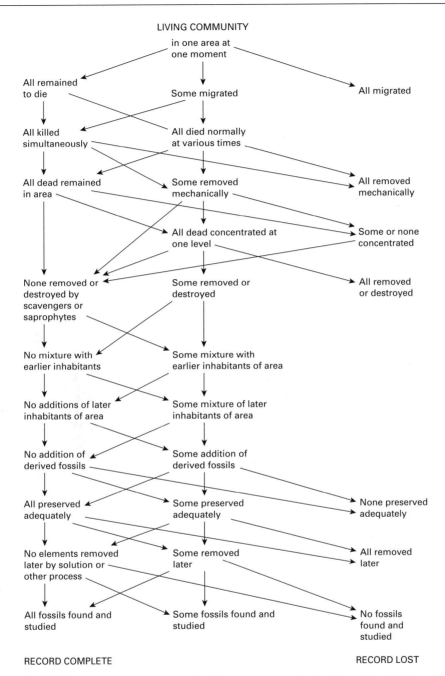

LIVING COMMUNITY
in one area at
one moment

All remained
to die

Some migrated

All migrated

All killed
simultaneously

All died normally
at various times

All dead remained
in area

Some removed
mechanically

All removed
mechanically

All dead concentrated at
one level

Some or none
concentrated

None removed or
destroyed by
scavengers or
saprophytes

Some removed or
destroyed

All removed
or destroyed

No mixture with
earlier inhabitants

Some mixture with
earlier inhabitants of area

No additions of later
inhabitants of area

Some mixture of later
inhabitants of area

No addition of
derived fossils

Some addition of
derived fossils

All preserved
adequately

Some preserved
adequately

None preserved
adequately

No elements removed
later by solution or
other process

Some removed
later

All removed
later

All fossils found and
studied

Some fossils found and
studied

No fossils
found and
studied

RECORD COMPLETE

RECORD LOST

Fig. 3.3 Aspects in the taphonomy of a living community. Adapted from Ager (1963).

Box 3.1 Experimental taphonomy

Experimental taphonomy is not new, and geologists have long been observing what happens to dead animals and plants on modern intertidal areas and, more recently, from submersibles. Or what happens to shells in a tidal channel, flume or wave tank. The subject has been reinvigorated by a better understanding of the processes of decay, particularly at the microbial level. Thus by studying the decay of the lancelet (probably the closest living animal to a conodont animal, Section 3.3.4) in a tank, it is possible to see how the skin decays, and how the muscles shrink, thus making it possible to explain with greater confidence some of the features present in the fossils.

Source: Briggs (1995)

Table 3.1 A scale of skeletal dissociation.

1 Sheathed and spiculate skeletons, where tissues loosely held together, e.g. holothurians, some sponges, alcyonarian corals.
2 Segmented skeletons where elements held together by muscle or ligament, e.g. crinoids, starfish, brittlestars, echinoids with imbricated plates, echinoid spines, small fish.
3 Bivalve skeletons without interlocking teeth and sockets, e.g. razor shells.
4 Bivalve skeletons with interlocking teeth and sockets, e.g. cockles, *Trigonia*, terebratulids.
5 Massive skeletons held together by muscles and ligaments, e.g. large vertebrates.

Taphonomy can also be used to distinguish between assemblages without necessarily identifying the taxa – a useful tool in facies analysis. Taphonomic processes are often considered to result only in loss of information. This is emphasized by the usual (more than 60%) loss of the soft-bodied (non-preservable) part of the biota. But there are a number of advantageous effects such as enhancement of growth banding in corals, anatomical preservation of plants, enhancement of trace fossils, and also time averaging and taphonomic feedback (Box 3.2).

Colour patterning is occasionally preserved. Preservation depends on the composition of the original pigments and the mineralogy and on the site of emplacement of the pigments. It is not uncommon in fossil insects; it is also found in molluscs when the pigment is preserved within the more stable calcitic shell layer (e.g. the gastropod *Theodoxus*).

The mode of preservation should always be considered, perhaps to establish the shape of the aperture of a particular ammonite, perhaps to establish the nature of the stomata of a particular leaf taxon, or perhaps to determine the original mineralogy of a shell. On inspection of any fossil, one has to keep asking the question: what additional information is needed and how can it be obtained?

There are four basic modes of preservation of animals and plants: the original material, replacement of the original material, impressions (moulds) of either of these, and encrustation. It is the combinations of these modes, together with the complexities of the organism and associated skeleton, that so often makes it difficult to elucidate how the organism was preserved. But it is absolutely essential to understand the modes of preservation before any further investigation can be pursued. How the sediment compacted, how the fossil accommodated this compaction, and whether or not compaction was arrested by early lithification, as with a concretion: all these are factors that may affect any fossil group.

Skeletal breakdown on death

In any study of skeletal material, consideration should be given to (a) how readily the skeleton may dissociate on death, and (b) how readily fragile skeletons (e.g. fragile corals and calcareous algae) may be attacked by bioerosion and lose their identity. For each site, attempt to determine a scale of dissociation (Table 3.1). Dissociation is determined by the readiness of the tissues holding elements together to break down, and the way that elements may interlock. Bioerosion is a major factor in tropical environments and may radically alter the make-up of the fossil sediment.

Buildups and stratiform skeletal accumulations can suffer wholesale destruction: loss of the vagile biota, addition of washed-in biota, bioerosion, algal encrustation and diagenesis (Box 1.1). The extent of this can be gauged by exploring a modern reef and then comparing it with a Pleistocene reef. (Due to sea-level changes, there are many sites where both are in close proximity, e.g. Caribbean islands.) Fig. 2.7 gives an indication of the problems in sections of a modern algal buildup in high and moderate energy settings. The more substantial the skeleton, the more pervasive the bioerosion (borings on all scales). Thinner and more delicate frameworks can be completely destroyed by burrowing crustaceans, making interpretation difficult.

Box 3.2 Three contrasting temperate intertidal sites in South Wales

Gower in South Wales (51°30′ N) is on the north side of the Bristol Channel. The tidal range is up to 8.6 m. Salinities on the open beaches are normal, and offshore water temperature is 16.5 °C (August) and 7.5 °C (February). A visit to three contrasting beaches was made when low water was about 13.00 hours, in order to demonstrate the differences in sediment, sedimentary structures and biota, and ecological controls on the biota and the taphonomy.

Rhossili beach (Figs 3.4 and 3.5) is exposed to the prevailing south-west wind and Atlantic storms. At the high-water mark there is a line of debris left by the receding spring tide: shells, arthropod carapaces, driftwood (with attached epiplankton), seaweed, dog-fish and whelk egg cases, pebbles, fish remains, bits of sponge, bryozoans and the inevitable plastic and other garbage. Many of the shells will have come from just offshore (*Donax*) or from the nearby rocky coast

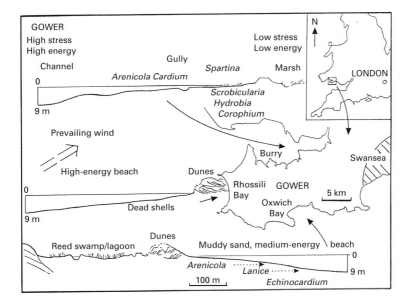

Fig. 3.4 Distribution of biota at three beaches in South Wales: a high-energy beach (Rhossili Bay) facing into prevailing wind with fauna living only at low water; a medium-energy beach (Oxwich Bay) with *Arenicola* flats, dune bar and backswamp/lagoon, and an estuary (Burry) with tidal channel and sand and mud flat. In the estuary, energy (stress) decreases towards the mud zone at high water, where the shoreline is prograding due to colonization by *Spartina townsendii*.

Fig. 3.5 Rhossili beach: looking north at low water on a calm day.

Box 3.2 (cont'd)

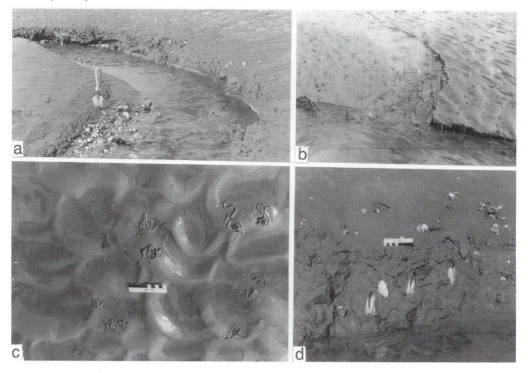

Fig. 3.6 (a) Burry estuary, tidal channel, looking downstream, with imbricated valves of *Cerastoderma* and *Scro-bicularia*; (b) *Lanice* (sand mason) tubes being eroded in a small ebb-tide channel, Belmullet, Mayo, Eire; (c) Penc-lawdd (Burry), lower sand flats with sand ripples and castings of *Arenicola marina* (lugworm); (d) *Mya* shells being eroded from sediment at outside of meander bend of small channel, Penclawdd, Burry. Scale lengths = 0.1 m.

(*Mytilus, Buccinum, Natica, Littorina*). Occasional land shells have come from the dunes behind the beach. The surface of the sand of the upper shore is smooth except for primary current lineation and traces of anti-dunes. At times of high wave intensity, the middle shore is thrown into sand waves, which are only in-completely levelled as the tide recedes. Digging shows no sign of life in the sand and, close to high water, dis-plays undisturbed beach lamination. Towards low-tide level, tubes of the sand mason *Lanice* are conspicuous, and if the tide is very low, one can spot the small openings of rapidly burrowing razor shells (*Ensis*).

In contrast, the beach facing to the south-east at Oxwich has a different aspect. The lower beach is covered by ripples, mostly somewhat degraded, and lugworm castings (Fig. 3.6). Digging reveals no lam-ination. Towards low-water mark *Lanice* (Fig. 3.6) may be abundant, but areas where this is absent may show the small inhalant opening of the potato urchin (*Echinocardium*). With care it is possible to see an emerging tube foot maintaining the tube. Digging will not reveal the 'orange-peel' layers formed by the animal's forward progression, they only show up in fossil burrows by diagenetic enhancement.

At Penclawdd the salt marsh grades into mud and sand flats, which are traversed by dendritic tidal channels (contrast with the drainage system formed in tropical mangrove environments). Here energy increases away from the shoreline and the upper flats are very muddy. Grazing *Hydrobia* are present in high numbers. There are two types of conspicuous trace: paired openings of the amphipod *Corophium* with its *Diplocraterion*-like burrow (Fig. 8.19), and the star trace above *Scrobicularia*, an infaunal deposit-feeding bivalve. On the sandier flats, scraping the sur-face with a fork shows many cockles (*Cerastoderma*). Annual growth and disturbance rings are clear. The small tidal channels are often floored with a layer of cockle shells (Fig. 3.6) and it is easier to walk along the channel than across the mudflat! Towards the main channel, evidence of storm degradation is common on the sandflats (Fig. 3.6). Ripple troughs often display abundant echinoid spines washed into the estuary.

Discussion might concentrate on the main ecologi-cal controls operating on the biota at each site, espe-cially on the role of the energy regime.

Guides: Guides to seashore life are widely available from museums, marine stations and bookshops

Some definitions

Impression Impression includes the surface(s) between fossil and sediment, or for encrusters, between skeleton and skeleton, or between skeleton and soft tissue (bioimmuration), as when an oyster encrusts a soft bryozoan.

Replacement Replacement includes calcitization; it is a change of mineral into itself or a polymorph when the gross composition remains essentially constant.

Natural cast Natural cast is the form that can be released from a (natural) mould. For example, a mould that results from dissolution of a shell in sandstone may be infilled naturally; decay of the pith may result in the cavity becoming filled to form a cast of the pith cavity.

3.2 Plant fossils

Plant remains vary greatly in their likelihood of entering the geological record. Spores and pollen, produced in vast numbers, are widely dispersed and most are highly resistant to decay, but may be produced seasonally or only once during the plant's life (Palaeozoic lycopods, bamboo). Leaves may be shed or blown from branches but still attached to twigs. Flowers may be seasonal and are always delicate. Timber may be transported far out to sea before becoming waterlogged. Roots have the highest preservation potential. Plant stratinomy (Spicer, 1991) is too complex to be considered in detail here, but the plants on a bedding surface are primarily the result of stratinomic processes. Fragile material is unlikely to have been transported far.

3.2.1 Vascular plants

Vascular plants exhibit three main modes of preservation. Each reflects the varying preservation potential (Table 3.2) of the different tissues making up the plant. Most of the problems that one experiences when trying to understand exactly what is represented by a particular plant fossil, are due to the often complex organization of the plant internally (Fig. 3.7), and between root, rhizome, stem, bud, leaf, flower, cone, seed, etc. Effects of compaction during peat formation, and the way in which the sediment splits, have also to be considered. Buried plant material may decay quite slowly so that the processes of decay, sedimentation and compaction are more or less concurrent. A Carboniferous log, or piece of peat, will have undergone compaction, probably by a factor of 6 or greater. Logs in a typical Upper Palaeozoic fluvial conglomerate can be compressed to slivers, albeit cm–dm across, and originally of similar thickness. But similar post-Palaeozoic lithologies generally display thicker logs because of the woodier stems prevalent from the Mesozoic onwards.

Impression

The quality of an impression (Fig. 3.8) depends much on the grain size of the sediment. Some leaf material may still be present as coaly remains or as carbonaceous upper and lower cuticles (if burial and temperature

Table 3.2 Plant fossil preservation modes, and relative resistance of cells to decay.[a]

Observations	PRESERVATION MODES Mode of preservation	Implications
Absence of carbonaceous, or coaly film	Impression	Only surface morphology available
Carbonaceous/coaly films	Compression ± impressions	Depending on thermal grade, cuticle, etc., may be prepared out
(With hand lens) cell details evident	Anatomical preservation (charcoal, silica, pyrite, carbonate)	Cell form and organization available
Associated with concretion	Any of the above	Minimal compactional distortion
	RELATIVE RESISTANCE TO DECAY	
LOW		HIGH
Cellulose	Lignified cells	Cuticle
e.g. parenchyma		e.g. pollen
		walls of spores

[a] This is a general model; exceptions will depend of the bacteria/fungi responsible for degradation.

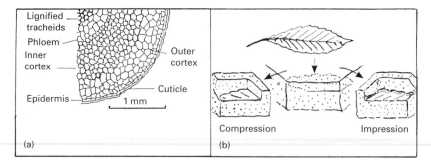

Fig. 3.7 (a) The early vascular plant *Rhynia* shows central lignified tracheids (protoxylem and metaxylem) then phloem surrounded by loosely organized parenchymal cells of inner and outer cortex and flattened cells of epidermis with cuticle (Rhynie, Scotland, Lower Devonian). (b) Leaf burial leading to compression and impression fossils.

Fig. 3.8 (a) *Alethopteris* as compression fossil in ironstone concretion; because the concretion formed at an early stage, the three-dimensionality is largely preserved (Westphalian). (b) Sandstone cast of decorticated lycopod root *Stigmaria* with rootlets, sometimes known as '*Knorria*' (Upper Devonian, North Devon, England). (c) Silicified cone of monkey-puzzle tree *Araucaria*, exterior and in longitudinal section (Jurassic, Patagonia). (d) *Stigmaria* (lycopod root); the inner cortex rotted and the space became filled with sediment, but the xylem rotted later and remained unfilled; the outer cortex is largely compressed; note the traces of rootlets entering the xylem cylinder at rays (Westphalian). Scale bars = 10 mm.

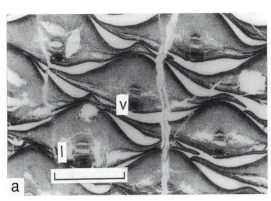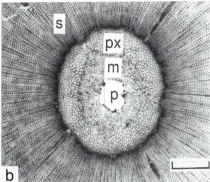

Fig. 3.9 Lepidophlois kansunum, Pennsylvanian, Kansas (peels). (a) Leaf cushion, ligule pit (l), vascular trace (v) and lateral ventilating tissue; scale bar = 10 mm. (b) Inner part of stele with pith (p) (largely lost), metaxylem (m), protoxylem (px) and secondary xylem (s); scale bar = 2 mm.

have been sufficient), but generally will have oxidized away. Sometimes a thin bacterially induced encrustation may have coated the leaf just before burial, enhancing the detail. More complex impressions result from a stem that is decorticated (has lost its bark, Fig. 3.8) or was already partly rotten before burial. Hollow stems, e.g. horsetails, can also become sediment-filled (with sand or compactable mud). With leaves, latex pulls can be taken from the impression of the lower cuticle (bearing stomata) for later examination under an electron microscope. Tufa often provides good impressions of leaves, stems and seeds; but irregular calcareous crusts on stems that probably formed in a desiccating environment, are generally unhelpful.

Compression

In compressions (Figs 3.8 and 3.9) of leaves, all that remains is coaly matter (*phytoleim*) between the superimposed cuticles. Generally, the sediment splits between one cuticle and the sediment impression. It may then be possible, if they are not too carbonized, to lift off the cuticles and prepare them in the laboratory. If the leaf had strongly curved margins, or was of complex shape, the sediment may not split readily to display a complete surface. Inspect a broken edge to detect the three-dimensional form, and the way the sediment splits. It is important to collect part and counterpart. Compressed megaspores may be readily seen in late Palaeozoic durainous coals.

Anatomical preservation

Anatomical preservation relates to the three-dimensional preservation of the cellular structure of the plant. However, decay and some compaction may have affected the plant before initiation of the processes leading to anatomical preservation.

Fossil *charcoal* (*fusain*) formed by rapid but incomplete combustion, in partial lack of oxygen, and in the fossil record can be attributed to lightning strike, wildfire and volcanic action. It is a strong but brittle material. Twigs, logs and even leaves, flowers and pollen can be charcoalized. Probably most fossil charcoal formed on the plant in its living position. Fusain is a common component of fluvial, deltaic and shallow-marine sediments. The sooty fusain of coal represents pieces of charcoal that entered the swamp peat.

Stabilization by minerals (e.g. calcium carbonate, silica, pyrite) occurs in the walls of plant cells or as minerals crystallizing within the cells (permineralization). This is the most valuable mode of preservation for the palaeobotanist. If the cell wall is still present as organic material, and has not been fully replaced (petrifaction), then acetate peels can be made by etching a polished surface with a dilute acid to form relief. The dead cells of wood (xylem) lend themselves to conducting fluids, and indeed, in volcanic regions (e.g. Mount St Helens WA) the wood is often partly silicified (like a plumbing system becoming furred) during the life of the plant. Rarely, cells that were once living have also become permineralized with the preservation of cytologic structures. Silicification is also found in sandy facies where silica is the main cementing mineral. But there are exceptions, with silicified plants occurring in carbonate sediments (e.g. Purbeckian of southern England).

Calcification, silicification and pyrite formation can be so rampant as to destroy much of the anatomical detail and what, in the field, might seem promising material may be disappointing in thin section. In part, this is attributable to early decay. Although best known from the calcified peats (coal balls) of the Silesian (Pennsylvanian), calcification is widely distributed

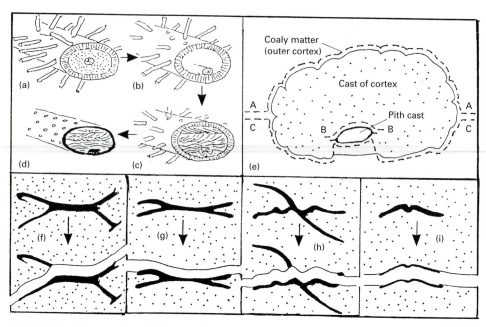

Fig. 3.10 (a–d) Preservation of *Stigmaria* (lycopod root): (a) in life with central stele (wood), surrounded by inner cortex (thin-walled cells) and outer cortex (thick-walled cells); (b) death, and decay of inner cortex, collapse of stele to floor; (c) space filled with cross-laminated sand; (d) stele and outer cortex compacted to compression state. (e) Stigmarian axis as in (d) with possible fracture planes: (A–A) passing through outer surface of axis, (B–B) passing over compressed stele and (C–C) exposing outer surface of axis and then jumping to expose surface of stele compression. (f–h) Lycopod leaves (*Cyperites carinatus*) as compression fossils in mudstone (but still exhibiting some degree of three-dimensionality when viewed in vertical section) with possible fracture planes yielding two-dimensional fossils. (i) Two-dimensional type (*Lepidodendron wortheni*) and fracture plane: the illustrated specimen shows incomplete compression, i.e. the fracture surface has passed *through* compression. Adapted from Rex (1983, 1985; Rex and Chaloner, 1983).

in associated volcanic facies. Keep a lookout for calcified twigs or roots (Fig. 5.14) in non-marine facies. Also be on the lookout for material for growth ring analysis. Thicker-walled late growth is generally the better preserved. A perplexing question is why coal balls are unknown from post-Permian strata.

In sandy or silty sediment, pyrite may nucleate on plant stems and seeds, or within woody cells. A stem may pass through a pyritic concretion. This type of preservation may look poor, but under the scanning electron microscope, details may be present. Similarly, limonite or phosphate may be present.

Intergrading

Impression, compression and charcoalified (fusainized) plant fossils may be associated with concretions (Fig. 3.8). The three preservation modes may intergrade. For instance, charcoal may be incorporated into a coal ball, or a stem in a coal ball may continue beyond the coal ball as a compression fossil. Logs of

wood may have become partly compressed before silicification (or calcification). Wood tends to be compressed parallel to the cell files. Autochthonous roots, upright stems and trunks are generally preserved as impressions of the bark, and compressions of what tissues remained (mainly xylem) after decay and infilling of the space by sand and mud. In lycopods (Fig. 3.10) the woody outer cortex may become compressed, and the small woody stele is often displaced. Particularly difficult to unravel in fossil plants are the different appearances that can occur between different modes of preservation of the same species. Even an exceptionally well-preserved root (Fig. 5.14) cannot be identified because botanists do not always detail the morphology of extant roots.

Most aerial parts of fossil plants are found as fragmented and dispersed remains. How can these remains be reassembled as a reconstruction of the original plant? Palaeobotany is much concerned with this detective work, and it is important to realize, in this respect, the significance of any material one comes across in the field. There are three methods of reconstruction:

- Locating material where, by good chance, two or more parts (organs) are joined, e.g. root and stem, cone with pollen and dispersed pollen, leaves and stem, leaves and reproductive organs. This is the only positive method, and it depends mainly on field observation.
- Identifying particular anatomical features that are common to two or more organs. Frequently cited characters are particular glands, type of stomata, cell outline – features that cannot readily be identified in the field.
- Commonly occurring associations of organs (e.g. on bedding surfaces) that would be difficult to explain if not from the same type of plant.

Bockelie (1994) has made a useful preliminary study of plant roots in core. Roots have high environmental significance, and some allostratigraphic relationships are shown in Fig. 8.28. But it is seldom that roots apparently in life position can be identified. The distinction between roots and burrows (Section 4.1.2) does not always work! In sandy cores, often all that is preserved is a carbonaceous film associated with sediment that has infiltrated primary spaces or that remains following tissue decay. For any succession describe the root morphology in as much detail as possible (attitude, branching, root, growth rings, nature of fill), and note relationships with other fossils.

3.2.2 Coal and oil shale

Coal is described in terms of rank (thermal grade, Fig. 3.11) and type (composition). The end members (anthracite and peat) can be fairly readily identified in the field. Very bright, splintery coal is likely to be of high rank (low-volatile bituminous coal to anthracite). The tectonic and sedimentary setting of the site, and the grade of the associated mudrocks, will provide a clue to rank. Is there a root bed present? Is the coal autochthonous or drifted? How 'pure' is the coal? It may be no more than a carbonaceous mudstone. Try to identify the components of a bituminous coal:

- *Vitrain*, representing coalified logs or pieces of bark, is glassy (vitreous) and usually closely jointed.
- *Fusain* generally forms lenses and has a silky sheen, but is soft and leaves a black mark.
- *Durain* is dull and tough, with megaspores often evident in Palaeozoic coals.
- *Clarain* (attrital coal) is finely laminated vitrain and durain.

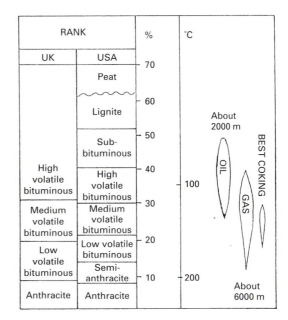

Fig. 3.11 Coal rank and percentage of volatiles: relation to oil and gas generation and coking potential. Adapted from Stach (1982).

This classification of coal is tedious to follow for logging. Try using mm, cm or dm thicknesses of vitrain, fusain or attrital coal. Further detail can be introduced by making estimates of the ratio vitrain/ attrital. Try to get an indication of the overall vitrain/ fusain ratio and the percentage of mud present.

Oil shale and cannel coal have formed in aquatic anaerobic environments. Cannel coal (dull with conchoidal fracture) represents drifted, finely divided terrestrial plant material. It is often found at the top of a coal, following a flooding event. Oil shales, and their principal components, are torbanite (lacustrine) with *Botryococcus* and allied algae; lamosite (lacustrine) with planktonic algae; tasmanite (marine) with *Tasmanites*, and marinite with various planktonic marine algae (Hutton, 1986).

3.2.3 Calcareous algae and stromatolites

The undulose, arched or planar laminations of stromatolites (Fig. 3.12) are generally readily recognized in the field. The laminations mostly reflect discontinuities in formation (e.g. storm action), but short (e.g. diurnal) periodicities are also recognized. Nodular stromatolites (oncolites) are distinguished by their irregular lamination around a shell or lithic clast. The clotted texture of thrombolites is not easy to prove in the field.

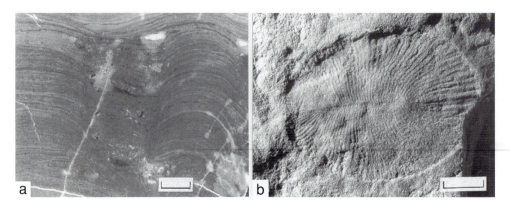

Fig. 3.12 (a) Stromatolite *Kussiella kussiensis* (Proterozoic, early Riphean, southern Urals). (b) *Dickinsonia* (late Proterozoic, Ediacarian, South Australia) on sandstone sole, as concave hyporelief. Scale bars = 10 mm.

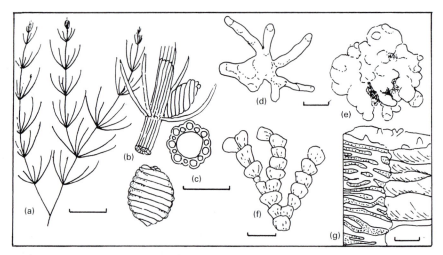

Fig. 3.13 Calcareous algae. (a–c) Charaphyte green algae (Silurian to Recent): (a) form of living plant *Chara* (scale bar = 10 mm); (b) detail of stem, branches and a gyrogonite (oogonia); (c) gyrogonite in side view and in cross-section; both views are seen in rock (scale bar = 1 mm). (d, e) Two modern thalli for calcareous red algae *Lithothamnion* (Carboniferous to Recent): (d) free-living growth form of overall spheroidal shape; (e) massive spheroidal form due to constant 'pruning' in an active depositional environment (scale bar = 10 mm). (f) Calcareous green alga *Halimeda* (Cretaceous to Recent) with calcified thallus; this readily breaks into segments and can make a significant contribution to the sediment, as may be seen on Caribbean beaches today (scale bar = 10 mm). (g) Growth manner of modern crustose red coralline alga (*Mesophyllum*); stippled areas represent intercrustose matrix; many Palaeozoic solenopores exhibit similar growth morphology (scale bar = 5 mm). Adapted from Bosence (1983).

Growth form is probably the best indicator of red, calcified algae with forms corresponding to the shapes of most types of breakfast cereal. It may just be possible to make out, with a hand lens, the cellular structure of a solenopore, or the larger, reproductive cells present in more advanced rhodophytes. Some rhodophytes will probably have an irregular, knobbly outline attributable to truncation and branching (Fig. 3.13).

Although the stem of the green alga *Chara* sensu lato (Stonewort) may become encrusted, it is the calcified walls of the cells enclosing the oogonia, the gyrogenites (often replaced by silica or seen as moulds), that are so useful as environmental indicators (essentially freshwater). Use a hand lens to identify the spiral cells. Marine, calcareous (aragonite) green algae such as *Halimeda* readily break up into segments, as does *Corallina* (articulated red alga); see Fig. 3.13.

3.3 Animal fossils

3.3.1 Soft tissue preservation

Soft tissue preservation is rare (Section 5.3). In the field consider what types of sediment may yield information on such material. There are five modes of preservation:

- Encasement in decay-inhibiting (or delaying) material; for example, scorpions in Carboniferous coal (peat), insects in amber, where desiccation will have been a factor.
- Encrustation of soft tissue (e.g. hydroids, algae) by skeletal organisms to leave a mould of the soft tissue (Box 3.3), but which may be deformed in the process.
- Casts or impressions of the tissues, where the sediment was stabilized by organic mats prior to decay, as in the late Precambrian Pound Quartzite (Fig. 3.12).
- Replacement of the tissue by minerals (clay, pyrite, silica, phosphate, carbonate) where precipitation was induced by bacterial action. Vertebrates and arthropods are associated with high levels of phosphate and the finest details of muscle fibres can be preserved by phosphatization. But this is only evident under electron microscopy. (Supposed fossilized blood corpuscles in the bones of dinosaurs are diagenetic pyrite framboids – microscopic pseudofossils.)
- Mineral coats, biologically induced to leave cellular impressions where tissues are outlined by a mineral.

Microbial mats are now recognized as perhaps the main influence on soft tissue preservation (Wilby *et al.*, 1996; Wilby and Briggs, 1997) by inhibiting diffusion of phosphorus from the sediment/organism into the water column. They can be identified in the field, thus anticipating soft tissue preservation, by irregularly lenticular and wispy layers of thickness a few millimetres.

Evidence of bioturbation, scavenging and early reworking almost always excludes soft tissue preservation. Although transport *per se* prior to deposition does not appear to be a major factor, aerobic decay during transport can be important. In general, increasing depth of burial (Fig. 3.14) leads to deoxygenation and then to anaerobic decay associated with manganese, nitrate, iron and sulphate reduction, methanogenesis and decarboxylation, which can promote the formation of pyrite, carbonate or phosphate.

Box 3.3 Bioimmuration

The poor representation in the fossil record of the soft-bodied biota (most annelids, many coelenterates and echinoderms) is much emphasized by biologists and geologists alike. Not only is the evolutionary history of these groups missing, but they must have played as important an ecological role in the past as they do today. Leads that can be found to close this gap are welcome.

Bioimmuration results from the rapid overgrowth by fast-growing organisms, such as oysters and serpulid worms, of sessile, soft-bodied animals. Their form then becomes moulded by the encroaching shell, though may suffer distortion in the process. Jon Todd completed his PhD on the topic and has demonstrated just how much information on soft-bodied animals can be revealed. He and colleagues have described and illustrated soft-bodied bryozoans, hydroids, algae on the surface of Jurassic and Cretaceous oysters – groups that have otherwise left little or no fossil record. The setae of brachiopods, the delicate hemichordate *Rhabdopleura* and goosenecked barnacles are among other finds. Bioimmuration, and an allied process, *epibiont shadowing*, are processes that take place under *normal* conditions, rather than in the special conditions of an anaerobic substrate (Section 3.3.1). Search the underside of encrusters, and between encrusters.

Sources: Taylor (1990), Taylor and Todd (1990)

These processes continue downwards until microorganism activity ceases. Thus, lack of oxygen does not prevent decay, but only retards it. To at least hasten a potential fossil entering the anaerobic zone, it is necessary to either entomb it in a decay-inhibiting environment (e.g. amber or peat), or bury it quickly, e.g. autoburial, as when a dead nektonic carcass nosedives into very soft mud. The fossil (and subsequent concretion) will then be oblique to the stratification.

There is some evidence that in clay-rich sediments all soft tissues may be bacteriogenically affected, but it is only when a concretion forms around the potential fossil is it likely to be preserved. Some concretions do have the fossils irregularly positioned, but the outlines of many give an indication of what is to be found inside. Decay resistance must be viewed on a geological timescale. Permafrost, peat bogs, tarpits and resin provide exciting insights and fossil DNA!

In freshwater sediments, *pyritization* of soft tissues is relatively unimportant due to relatively low levels of sulphate in the water. In marine sediments with

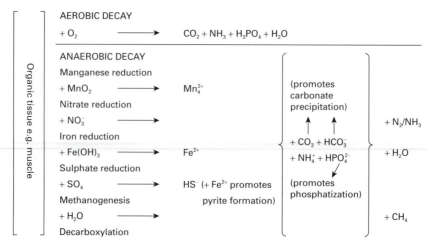

Fig. 3.14 Ideal stratification of bacterial reduction zones. Nitrate reduction and methano-genesis dominate in freshwater environments, sulphate reduction and methanogenesis dominate in marine environments Note that iron (Fe^{2+}) may be available during the whole period. Adapted from several sources, including Allison (1988) and Eglington *et al.* (1985).

sufficient iron, pyritization is favoured by high burial rate. With rapid burial, more reactive (less decayed) organic matter is added to the sediment, leading to a rapid uptake of sulphate ions from pore water and therefore increased diffusion of sulphate ions from the overlying seawater. In euxinic conditions, organic carbon is generally too high to favour good pyritization and pyrite is disseminated.

The *phosphatization* of soft tissue (Briggs and Wilby, 1996) is generally associated with low burial rate (but without reworking) and high organic input. Phosphatic materials (bone, arthropod carapace) often act as nuclei, but the best phosphatization seems to have taken place, or at least to have been initiated, at the sediment–water interface (Martill, 1988; see also Section 5.5).

Early *carbonate encasement*, where calcium carbonate was precipitated as concretions, is favoured by high organic input and burial rate. Where terrigenous input is low in marine or freshwater settings, rapid carbonate precipitation is favoured by high carbon dioxide production by algal and cyanobacterial mats, e.g. the bedded limestones of Solnhofen (Upper Jurassic, Germany) and the Green River Formation (Eocene, United States) where, however, fossils always seem to be compressed.

3.3.2 Univalves and groups with skeletons that remain more or less intact

Graptolites As with ammonoids (see below), the mode of preservation for graptolites in a succession

of mudrocks may change appreciably from bed to bed, indicating small variations in facies. Graptolites had a pliable organic (collagenous) skeleton. Three-dimensional preservation in early formed concretions and certain limestones can be collected for later acid treatment, but is uncommon. The most common mode of preservation is as carbonized side-down (profile) compressions, or compressions where the organic skeleton has been replaced by white-weathering, phosphatic material (Fig. 3.15). Diplograptids, however, commonly landed on the sediment with one series of thecae being entombed aperture down (scalariform view). If the form of the thecae is clear, identification is generally possible. Locate material with discrete rather than crowded specimens and with the proximal morphology distinct. Take care not to misinterpret apparent branchings for *Cyrtograptus* or *Nemagraptus*. Commonly, pyrite filled the thecae at an early stage. Only internal moulds may remain, but they have the advantage of being undeformed.

Where mudrocks contain distal turbidites (striped mudstones), the graptolites therein tend to be richer and better preserved than in more oxidized sediment. Preservation was enhanced by rapid sedimentation.

Conularids Conularids are generally highly compressed in mudrocks. Three-dimensional preservation, with external and internal impressions, occurs in sandstones.

Archaeocyathids Archaeocyathids generally have a skeleton with distinct morphology (Fig. 3.15). The

Fig. 3.15 (a) Diplograptids with micritic infill, in micritic pebble in conglomerate; one specimen in scalariform view. (b) Archaeocyathid with skeleton slightly etched from matrix (Ajax mine, Flinders Ranges, South Australia, Lower Cambrian). (c) *Didymograptus* compressed as a white film (Lower Ordovician). (d) Fenestellid bryozoan, mainly displaying non-apertural side, but small area right (by sketch) with apertures (Dinantian, Lower Carboniferous, North Wales). Scale bars = 10 mm.

calcareous skeleton will now usually be of calcite or will have been silicified. It may have originally been Mg calcite. Geopetal sediment and sparry fill to chambers is common. Check for epitaxial algae (*Epiphyton*) as dark micritic clots on autochthonous material. Mouldic preservation should show evidence of regularly porous walls.

Sponges   It is exceptional for a fossil sponge to be found relatively perfect. The best preserved are those with fused mineralized spicules giving a rigid skeleton. (Peripheral, loosely bound spicules will still be missing.) Judging by the abundance of loose spicules in many sediments (e.g. the needle-like spicules in many Jurassic and Cretaceous rocks, or the ball-bearing-like dermal spicules (*Rhaxella*) in

the Upper Jurassic) sponges were much more widespread than is now apparent. Opaline spicules of demosponges and hexactinellids are readily dissolved. Look out for moulds (in sand and mudrocks), or carbonate replacement in limestones. Sclerosponges have a basal calcareous skeleton with embedded siliceous spicules. In fossil representatives (e.g. *Chaetetes*) the skeleton may now be entirely calcitic.

Rugose and tabulate corals   In sandstones, preservation is nearly always poor. If the skeleton is present, it will be almost certainly replaced. Sediment may have infilled the calyx and infiltrated more deeply, giving variable perfection to moulds. Preservation in mudrocks can be good, especially if they are somewhat bituminous. Compaction is common in

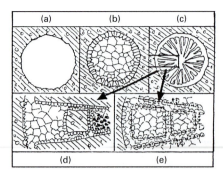

Fig. 3.16 Preservation modes of a scleractinian coral (aragonite): (a) void with impression of external morphology; (b) calcite-filled void, cast; (c) retention of septa by either (d) spar replacement between matrix-filled interseptal spaces or (e) calcitization of aragonitic septa (with distinct micritized wall to septa) in a phreatic environment. In (d) a shelter cavity was partly infilled by pellets. It may then be inferred that the wackestone in which the coral is embedded was originally pelletal. Magnification for (a–c) is approximately ×0.5–1.0.

wackestones and lime mudstones. Select material suitable for transverse and longitudinal sectioning. The manner in which colonial forms increase is important.

Scleractinian corals Due to the originally aragonitic skeleton, mouldic or replacement modes of preservation are widespread in scleractinian corals (Fig. 3.16). Neomorphic preservation (calcitization) is mostly associated with massive colonies (because they were less permeable). Colonial and solitary forms can be completely dissolved and the resulting cavity spar-filled. Preservation is generally poor in reefs, becoming better laterally in muddier facies. For full identification it is essential to have good preservation of the skeleton for transverse and longitudinal sectioning. Distinct banding is often evident, probably emphasizing original seasonal growth variation.

Bryozoans The stony bryozoans of the Palaeozoic had a calcitic skeleton and are like miniature tabulate corals, showing similar preservation modes. Preservation is often clear in mudstones with specimens weathering out. Good preservation of external morphology and then sectioning is necessary for close identification. Fenestrate bryozoans are typically exposed with the apertures (smaller than the fenestrules) facing down (Fig. 3.15), so it is necessary to search for material with apertures facing up. Slender, encrusting cyclostomes are easily damaged. In many cheilostomes the frontal surfaces include delicate sculpture, ovicells and avicularia, costae and spines. Some loss of detail

may have taken place because of abrasion or dissolution. Also replacement usually leads to loss of detail, though silicified bryozoans within hollow flints may be well preserved externally. Impressions of the external surface of Palaeozoic forms are of little significance.

Larger foraminifera Larger foraminifera (Box 6.1) are common at certain times and in certain facies and are generally well preserved in fresh rock. Fusulines (Upper Palaeozoic), alveolinids (Mesozoic onwards) and rotalids (e.g. *Nummulites* and orbitoids in the Tertiary) all had calcitic skeletons. In limestones the preservation is generally good. Individuals may weather out, or blocks may be collected for later slabbing. In sandstones and mudrocks, weathering may lead to dissolution. Break some specimens open to check that skeletal detail is present.

Stromatoporoids Stromatoporoids are a heterogeneous group. Break off a piece of rock and examine the fresh, wetted surface with a hand lens to spot the skeletal laminae and pillars (never present in stromatolites). Thin sections are required for identification.

Gastropods Most modern gastropods have an entirely aragonitic skeleton. The operculum, when present, is horny or aragonitic. The preservation mode is similar to ammonites but, being generally without septa, sediment readily enters the spiral shell. In bioclastic limestones and sandstones, and in oolites, preservation is generally as impressions of exteriors, infills (cores) or with replaced shell. With impressions, it may be more difficult to locate specimens with the earliest growth stages preserved and the terminal aperture intact. In mudrocks, compressed material can be difficult to determine. Spinose forms will need care in collecting.

Serpulid tubes Serpulid tubes are common, free or attached, and generally well preserved with the original skeleton, or as mouldic preservation. Many are encrusters but the host may often not be evident if it was aragonitic and has been dissolved. Massed tubes of serpulids can form substantial aggregations. Other annelids line or construct their tubes with a variety of materials (Fig. 8.30).

Scaphopods Scaphopods have always had an aragonitic skeleton (so far as is understood). They are uncommon as fossils until the Tertiary and then can often be found with original aragonite in mudrocks.

Externally shelled cephalopods Nautiloids, ammonites, goniatites and ceratites are among the

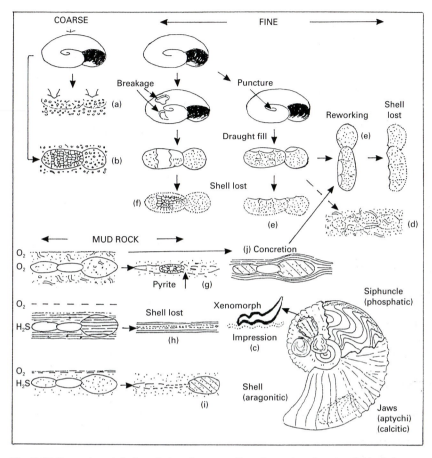

Fig. 3.17 General model of cephalopod preservation: for explanation see Table 3.4.

most common of macrofossils, readily recognized and often exquisitely preserved (Fig. 3.19, see p. 48). Evidence of death is occasionally evident as bite marks. Today *Nautilus* conches are dispersed by wave or current action far beyond their 'home waters'. The same is likely to have occurred with ancient cephalopods. Their taphonomy is astonishingly varied and complex. In particular, their chambered organization and, in many instances, their shape, means that local microenvironments, formed during diagenesis, are 'out of step' with the general diagenetic gradient (Fig. 3.17), e.g. pyrite forming only in the chambers. The different modes of preservation have important implications for palaeontology, stratigraphy and sedimentology.

Although there are many variables, identification can often be achieved using a few features:

- *Palaeozoic nautiloids*: form of conch, nature of siphuncle and cameral deposits, form of aperture.

- *Palaeozoic ammonoids*: suture line, perforate or imperforate umbilicus, form of conch and venter, lateral sculpture.
- *Mesozoic ammonoids*: form of conch and venter, ribbing and tuberculation, suture line.

For information on ontogeny, pyritized or phosphatized inner whorls are needed. Where reworked cephalopods are present, take care in making stratigraphic conclusions. Take advantage of any differences in preservation mode that may be present, which can provide information about the early diagenetic regime. The factors that seem to have most influenced the mode of preservation are shown in Fig. 3.18.

It is generally possible to establish by which of four methods the shell was infilled by sediment:

- Direct filling of body chamber.
- Damage to, or boring through outer wall, leading to filling of connected chambers.

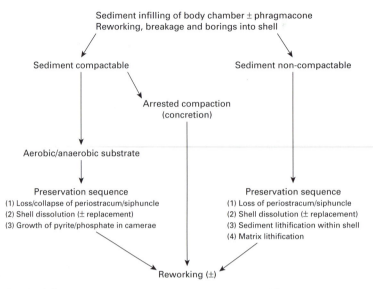

Fig. 3.18 Flow chart to illustrate ammonite preservational factors.

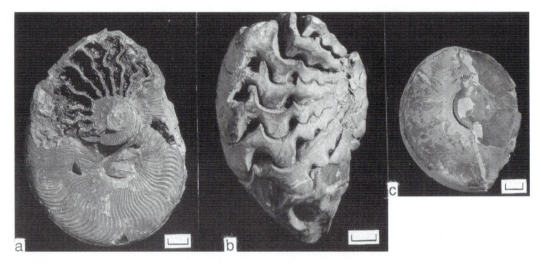

Fig. 3.19 (a) *Harpoceras* with some camerae filled with bitumen, within a concretion (Lower Jurassic). (b) Ceratitic ammonoid with draught fill and geopetal structure, shell subsequently dissolved (Middle Triassic, Germany). (c) *Harpoceras* in body chamber preservation, with rest of shell flattened (Lower Jurassic). Scale bars = 10 mm.

- By draught fill through septal necks, following damage or decay of siphuncle, thus allowing a current to flow through the shell and deposit sediment.
- Sediment entering, following partial dissolution of shell.

Other aspects that have influenced mode of preservation are as follows:

- Preburial encrustation by oysters, serpulids (see below, under xenomorphic preservation) and stromatolites.

- Attitude of shell on embedding, i.e. vertical or horizontal.
- Shell form could be involute with inner chambers protected or evolute with inner chambers unprotected; strong keel, e.g. *Asteroceras*. A rounded venter, e.g. *Nautilus*, may lead to irregular shell fracturing and to lateral spread under compaction.
- Pyritic films on outside (e.g. giving the limonitic film on many Chalk ammonoids).
- Growth of carbonate or pyrite in chambers or infilling of chambers by hydrocarbons (Fig. 3.19).

- The thinner shell of the first formed whorls, which may be more readily damaged.

In the Palaeozoic, many orthocones had a complex siphuncle and careful collecting is required to determine this. The great length of some orthocones means that a specimen might not be completely enclosed by a concretion. Note that pressure dissolution (stylolite) often leads to loss of information on suture line and sculpture.

The sedimentological and palaeontological implications of various preservation modes are shown in Table 3.3. There are four main preservation regimes; the following list is partly based on Seilacher (1971, 1976):

- Sandstones
- Oolites and bioclastic limestones
- Micritic limestones
- Mudrocks

In *sandstones*, ammonoids are generally found in mouldic preservation. In *oolites and bioclastic limestones* (Fig. 3.17a & b), compaction is not an important factor. But what is important is whether shell dissolution preceded (a) or succeeded (b) lithification. If shell dissolution preceded lithification, sediment more or less just dropped into the camerae as septae and outer shell dissolved. Sometimes all that remains is an impression of the trough form of the underside of the shell. (Such 'half-ammonites' can also be formed by later dissolution.) But if shell dissolution succeeded lithification, only the body chamber was filled with loose sediment, and the camerae (unless damaged) remained available for possible sparry fill. The shell was then replaced.

The *micritic limestones* (Fig. 3.17d–f) show a puzzling range of preservation modes. In Cretaceous chalk, since the substrate was often below the aragonite dissolution depth, ammonites (and gastropods and corals) are preserved only at specific horizons. Concentrations of original calcitic aptychi in Tethyan pelagic limestone are due to dissolution of the conch on the substrate. If the shell was draught-filled and buried, and dissolution took place but without lithification, then burrowers were able to pass through. The fill was subsequently deformed by compaction, and septal traces closed and lost. A composite mould may also result (d). In many sequences of condensed limestone (such as the Tethyan Ammonitico Rosso) and hardgrounds, reworking or, at least, re-exposure was common. If the draught fill was lithified before reworking, as in the Triassic Muschelkalk, the cephalopod was no more than a special type (e) of

intraformational clast (Fig. 3.19). In the micritic limestones of Solnhofen (Upper Jurassic), ammonite preservation resembles that of bituminous shales – compaction took place as the shell dissolved. (Note that draught filling of the phragmacone is usual, but not always possible to prove because of later compaction.) For phosphatic and glauconitic concretions, see Section 3.4.

In *mudrocks* (Fig. 3.17g–i) the lithification has been delayed so that compactional effects, growth of pyrite and phosphate, and the formation of concretions are the important factors. Pyrite that nucleated on the cameral walls hindered compaction, allowing preservation, especially of the inner whorls (g). While body chambers readily became filled by sediment, unless a concretion formed, compaction took place (h and i) (Fig. 3.19). Generally, the environment was too quiet for draught filling to occur. Cephalopods that came to rest on muddy substrates were natural 'islands' for settlement of hard-substrate oysters and serpulids. Since these are calcitic, attachment areas and xenomorphs (c) may be the only evidence. However, many oysters with impressions or xenomorphs of ammonites in mudrocks result from aragonite dissolution in the weathering zone.

Belemnites and other coleoids   Every geologist is familiar with belemnite guards, and the calcitic guard or rostrum with part of the aragonitic chambered phragmacone is also common. But the anterior extension of the phragmacone (pro-ostracum), which was not always completely mineralized, is uncommon. The phragmacone and pro-ostracum are nearly always crushed where unprotected by the rostrum. To find them with the ink sac and tentacles (with chitinous hooks) all together, it is so rare that composites have been made! The chitinous arm hooks may be spotted in clayey sediment with a hand lens.

3.3.3 Groups with skeletons that separate into two parts

This section largely concerns the bivalve shells: brachiopods, Bivalvia, ostracodes, branchiopods (Fig. 3.20).

Brachiopods   In Palaeozoic sandstone/mudstone/shale facies, representing mostly shallow marine environments, brachiopods are commonly found as impressions (moulds) of dissociated shells (Fig. 3.21). A minimum of four impressions are required for later casting (Appendix A): the outside and inside

Table 3.3 Observations and implications of ammonoid taphonomy.

Observations	Sedimentological implications	Palaeontological implications
OOLITES AND BIOCLASTIC LIMESTONES		
Shell lost ± impression of lower surface only Large bioclasts in infill	Early dissolution of shell	Loss of information of external sculpture and suture lines, ontogenetic information, body chamber and apertural form, cross-sectional form
Conch preserved (replaced), uncompressed; camerae spar-filled and/or draught-filled or, with damage to phragmacone, matrix-filled	Matrix lithified before shell dissolution	Good preservation of sculpture, growth lines and of suture lines (when shell removed)
MICRITIC LIMESTONES		
Attachment impression, xenomorphic features and/or aptychi (no aragonitic fossils)	Substrate below aragonite compensation depth	Loss of palaeontological and stratigraphical information
Bioturbated (and often deformed) internal/composite mould Sutures ± lost	Shell dissolution and plastic deformation Delayed lithification	Only gross features remain
Uncompacted body chamber and phragmacone as internal mould	Early lithification of fill	Loss of information on sculpture, and generally some loss of suture lines
Micritic fill to some chambers, others missing or spar-filled (i.e. local fill following damage to shell) Solnhofen mode (as for bituminous shales)	Early lithification of fill ± spar formation	Preservation variable, search for suitable material
MUDROCKS		
Bioturbated mudrock. Pyrite associated with ammonoid; early camerae with pyrite infill	Pyrite/organic content low, but local environments if sulphate reduction in camerae; aerobic substrate	Good preservation of inner whorls and possibility of ontogenetic studies Often difficult to locate outer whorls
Bituminous shales. High organic content; ammonoid completely flattened; aragonite dissolved ± thin periostracum film; pyrite disseminated in sediment	High compaction; anaerobic substrate	Identification difficult if crushed Sculpture distinct only when early fracture of body chamber has occurred Phragmacone dissolved rather than crushed Loss of suture line usual
Restricted mudrocks. Pyrite low to absent; aragonite shell generally present (or replaced), but crushed beer-mat preservation ± concretion (phosphatic or calcareous), body chamber generally evident (Fig. 3.19)	Aerobic layer just below substrate; intermediate organic content	Often possibilities for ontogenetic studies when inner whorls pyrite-lined (not sediment-filled) Body chamber concretion may not extend to aperture

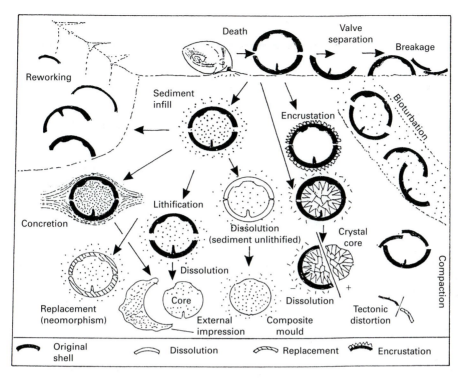

Fig. 3.20 General model for the taphonomy of a bivalve shell.

of each valve. Details around the pedicle area and dentition are particularly important. It may be necessary to collect several specimens to make up a complete impression of a brachial valve. Fine-grained to very fine-grained sandstones provide the best moulds. Avoid wet, weathered or argillaceous sandstones. It is generally possible to get an appreciation of a particular bed by inspection of the weathered edge. Weight of the blocks is also indicative. But some blocks that look promising from the outside will turn out to be blue-hearted (Fig. 3.21). Non-decalcified material can be collected for later acid decalcification.

Problems encountered in limestones are the welding of the shells to the matrix and the way pseudopunctate shells tend to split within the thickness of the shell (then displaying the pseudopunctae). Details of the umbonal area may be difficult to determine, and the rock seldom breaks to reveal internal structures completely. With joined valves, serial sectioning may be needed later. Many brachiopods have delicate spines and margins. Ensure that all such detail has been collected.

Mesozoic and Cenozoic brachiopods better resisted disarticulation, typically retain their original skeleton and are generally easily released from the sediment.

Inarticulate brachiopods with organic shells are often crushed and crumpled.

Bivalvia There is much variation in the ligamentation of bivalve molluscs and this is reflected in the readiness of the valves to dissociate. Bivalve molluscs show more complex diagenesis than brachiopods. Calcitic oysters are nearly always preserved with original mineralogy. In other bivalves it is the aragonite that is so prone to dissolution. For instance, in Chalk (Upper Cretaceous) *Inoceramus*, the nacreous inner layer, though relatively thicker, is often lost and all that remains is the thinner, outer (and often fractured) calcitic (fibrous) layer (Fig. 3.21). Compaction will have generally closed the chalk about the prismatic layer so that all trace of internal marking and muscle impressions is lost. But specimens can be found where the aragonitic layer has been replaced by calcite. In calcareous sediments, preservation can be better in allochthonous shell beds (which were cemented before aragonite dissolution) than in associated micrites (where the shells were leached before cementation). Full identification requires complete valves (or equivalent impressions) with dentition, muscle and ligament areas. Thin-shelled bivalves are liable to breakage and compaction.

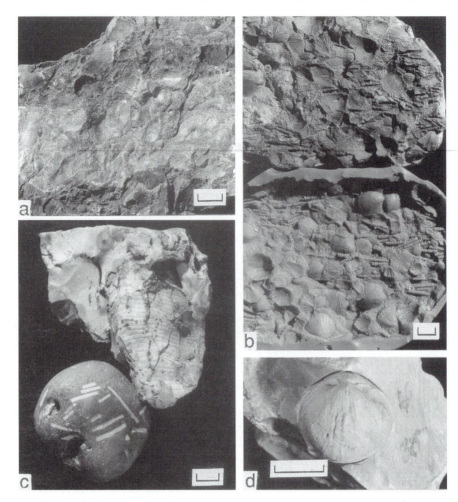

Fig. 3.21 (a) Partly decalcified calcareous sandstone with dissociated brachiopods: 'blue-heart' preservation (Ordovician). (b) Decalcified fine sandstone with moulds of orthid brachiopods and tentaculitids, together with plastic replica (Ordovician). (c) *Inoceramus* in flint pebbles: in the rounded specimen are fragments of the outer prismatic layer; in the other specimen the prismatic layer has been lost, revealing the surface between the prismatic layer and the inner nacreous layer, which was lost before formation of the flint (Upper Chalk). (d) Terebratulid brachiopod encased within and without by flint, shell dissolved (Upper Cretaceous, England). Scale bars = 10 mm.

Ostracodes and branchiopods   Ostracodes and branchiopods exhibit virtually the same preservation modes as brachiopods. It will generally be possible to release the valves from mudrocks and calcareous sediment. Mouldic preservation can be collected for later (latex) casting.

3.3.4 Groups with skeletons that separate into many parts

Echinodermata   Echinodermata (Fig. 3.22) have only one chance to be preserved complete and that means, in effect, burial in life position or very soon after death. Following death, no reworking (hydraulic

or bioturbation) is possible without dissociation taking place. Even without reworking, there is likely to be loss of the smaller skeletal elements such as pedicellaria, the smaller miliary spines of echinoids and pinnules of crinoids. Five grades of preservation can be distinguished for echinoderms. The following list is adapted from Smith (1984):

- Near-to-perfect preservation with, for instance, echinoid with spines, pedicellaria, lantern and apical disc intact. This does occur but has required exceptional conditions of smothering (storm, slump). No reworking will have taken place. If compaction has occurred (likely with the empty test), the oral and

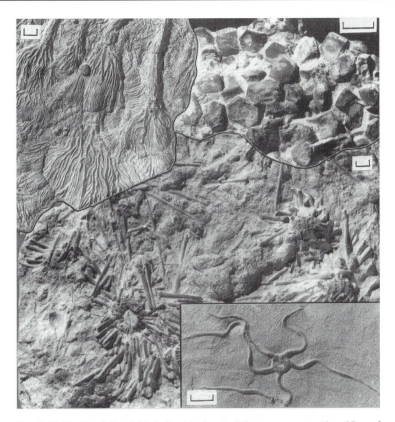

Fig. 3.22 Group of *Hemicidaris* in virtual autochthonous preservation. View of undersurface with oral-side-down test on shelly packstone. The group was overwhelmed by oolite (Bradford-on-Avon, England, Middle Jurassic) Scale bar = 10 mm. (Upper left) A group of *Pentacrinus* drawn out, possibly below a sunken log to which they had been attached when the log was floating (Lyme Regis, England, Lower Jurassic). (Upper right) Dissociated plates of *Palaechinus* (Dinantian) (northern England). (Lower right) *Palaeocoma* (brittlestar) on sole of storm event bed in autochthonous preservation (Dorset, England, Middle Lias).

aboral discs of echinoids will have collapsed first. Crinoid cups were generally stronger.

- Dissociated skeletons, where plates of the calyx are separated, but the stem and arms belonging to the particular calyx can be recognized adjacent to it. In echinoids the spines may be seen detached around the test. The stratinomical processes took place under quiet conditions.
- Good but incomplete preservation with loss of arms, and probably stem of crinoids, and loss of spines, apical disc and lantern in echinoids. This is the most common preservation mode of crinoids, echinoids and blastoids in collections.
- Preservation as abundant, dissociated, skeletal elements, except that a few crinoid ossicles may still be joined, for instance. It will be extremely difficult or impossible to reconstruct a whole individual. Nevertheless, identification to class or even genus may be possible.

- Small crinoids, brittlestars and holothurids readily break up. Their remains may be recognizable only on later microscopic examination of a washed sample but will provide important ecological information.

In the Palaeozoic many echinoids had an imbricate-plated test, and they are much less likely to be preserved intact than those with rigid tests. In general, epifaunal echinoderms in shallow-water environments have a lower preservation potential than groups that live infaunally or in quieter and aggrading environments. An exception may be found in the sand dollars with their strong, pillared test and deeply interpenetrating pegs and rods.

Most elements of an echinoderm skeleton are a single crystal of porous stereom. Syntaxial overgrowth is likely to occur in any sediment that is reasonably permeable leading to a readiness to break with calcite cleavage. (Preburial breakage is always irregular

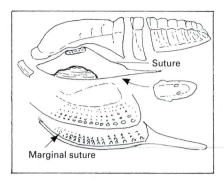

Fig. 3.23 Exploded trilobites to illustrate potential dissociation of exoskeleton. The trinucleid (below) has a marginal suture.

across the stereom.) In muddy sediment, the stereom may have been penetrated or remain virtually unfilled, though a thin film of iron sulphide is often present. Washings from Tertiary mudrocks yield spatangid spines looking as clean as any modern spine. Palaeozoic crinoid ossicles have been treated in the laboratory with hydrofluoric acid, which removed siliceous matter and replaced the stereom with fluorite. In sandy sediment, dissolution leads to moulds. Moulds are common in flint and chert.

Trilobita Preservation modes for trilobites (Fig. 3.23) are similar to those for brachiopods; the difference is the readiness with which the skeletal elements dissociated. Like brachiopods, it is generally difficult to obtain material from unweathered sandstone or siltstone. Complete specimens are not uncommon, preserved with the original calcite or as moulds. With dissociated material, try to collect the components to make up an 'individual' (though it may not be possible to determine the number of thoracic segments). If several moult stages are present, it is possible to evaluate ontogenetic stages.

Problems arise in collecting mouldic material; for instance, the components necessary to reconstruct the brim of a trinucleid. External impressions of both upper and lower parts are required: the lower part bears the genal spines. Note that the eyes of most lower and middle Cambrian trilobites are ringed by a suture and are generally separated from the free cheek, in contrast to stratigraphically younger trilobites. In fresh limestone, pygidia often seem to be more common than cranidia, but the rock tends to part less readily along cranidia. Compaction in mudrocks can pose problems.

Insects Insects are not so uncommon. If one considers how closely any non-marine environment

may be correlated with its insects, then a few insects should be very useful in the interpretation of ancient environments. Insects in amber (the exudative of gymnospermous wood, and equivalent to modern copal) are the best known. Amber is found generally as reworked pieces in terrigenous and shoreline sediments. Keep an eye open for amber in less likely facies, e.g. Tertiary flysch sediments and Cretaceous fluvial sediments. Older ambers are important because they can provide information on the evolution of the class alongside angiosperm evolution and diversification. In lacustrine and coal-bearing facies, insects occur carbonized or as impressions (often fragmented) in siltstones. Beetle elytra and cockroach wings are the commonest elements. Preservation of wings differs from leaves in that there is no compression of the membranes (since they are in contact), except along the veins (which may be pyritized). Silicified or calcified material is also known, as well as insect mines in leaves and wood, insect faeces, caddis fly cases, and pyritized pupae. Identification can be difficult, so seek expert help.

Crabs, shrimps, horseshoe crabs, eurypterids
Shrimp bed or lobster bed are common stratigraphic designations and nearly all are associated with fine-grained lithologies. Enclosure within nodules (commonly phosphatic) has enhanced preservation, and distinguishing between corpse and moult can be difficult: the corpse will have been attractive to scavengers. Fragments are also quite common in a variety of preservation modes, but are often indeterminable even to generic level, because of incompleteness. Nevertheless, even the impression of an elongate chela of a predatory decapod provides a useful bit of ecological information. Chelae are often strongly calcified, thereby raising their preservation potential. Small crustaceans in fine-grained limestones are generally compressed and their cuticle carbonized or replaced, e.g. by fluorapatite (white), silica or calcite. The cuticle may also be invested with pyrite, which may intrude the thickness of the cuticle.

Vertebrates Few of us are going to stumble on, or indeed locate, a large fossil vertebrate. If we do, we should get in touch with a local museum or vertebrate palaeontologist who has experience in extracting such material. Large bones weathering out of softer sediment often provide the first indication. When more complete skeletons are found, it is essential that the blocks are extracted with known relationships. More common are dissociated bones and fragments, loose scales and denticles, teeth and otoliths, more or less evenly distributed through the sediment, or hydraulically concentrated – bone beds and fissure fills. This material may be largely first-cycle, but also expect reworked material. Even so, bones and other vertebrate material may be rather dispersed, for example, in a conglomerate in which the main components are large logs. In this case consider concentrating the bones by sieving. Reworked material will probably have been polished and altered somewhat (pre-fossilization), especially by the infilling of pores and cavities. This leads to sharp, clean fractures in contrast to the irregular fracture of 'fresh' bone (autochthonous or parautochthonous). As with echinoderms, preservation of complete skeletons depends much on facies. Size is also a factor.

The upper and lower sides of an ichthyosaur carcass on mud may well have been decomposing under quite different chemical regimes – the upper side aerobic and the lower anaerobic. Professional collectors will always prepare from the underside, since this will be better preserved. If the water above the substrate was oxygenated then scavengers may have disrupted the skeleton, and encrusters (especially oysters) may have obscured it. In shelly layers, skeletons are likely to be somewhat dissociated because of reworking and/or hydraulic activity and scavenging. In bituminous mudrocks, isolated bones have probably dropped from decomposing, floating carcasses. Preservation within calcareous concretions will necessitate acid preparation in the laboratory. Often the skeleton extends beyond the concretion. Look out for unusual aspects, such as evidence of skin (which may be no more than a colour change or film beyond the skeleton); or an impression in sand of the detail of a footprint; or stomach contents (revealing diet, e.g. coleoid hooks in Mesozoic marine reptiles), or unborn juveniles.

Ostracoderm and other Devonian fish have large and often massive bony plates for part of their armour, which were highly resistant to attrition. They are a common component of fluvial gravel lags. Complete skeletons can be found in flood and slump deposits. But, under the acidic conditions of a peat bog, for instance, enamel is slowly decalcified, leaving only dentine. Similarly, bone can become decalcified and 'soft'. In organic-rich mudrocks, bone becomes compressed and distorted. The cartilagenous skeleton of sharks preserves less readily, though it may become replaced. Calcitic eggshells can be common in Mesozoic red beds.

Conodonts can be readily detected (using a hand lens) with various preservation modes: in white, where mudrock has been baked; as impressions in siliceous shales, and as centres to reduction spots (Section 7.7.5), or simply as small black 'teeth'! They can often be profuse.

3.4 Concretions

It is always a joy to find a complete fossil encased in a concretion (nodule) and several are illustrated in this chapter. The aspect that arouses most enthusiasm is the three-dimensional form of the scarcely compacted specimen, particularly if material from the adjacent sediment is strongly crushed. But there is a price to be paid – the difficulty of extracting the specimens! Always try and split a concretion parallel to stratification. Occasionally concretions are oblique to stratification (Section 3.3.1.) Keep counterparts, especially if the rock has split within the skeleton. In mudrocks the amount of compaction that the fossil has suffered will relate to the time of formation of the concretion. Septarian concretions can often be disappointing because of extensional cracking (which may also pass through the skeleton). The earlier the concretion formed, the more likely it is to have enclosed organic parts that are normally lost, e.g. soft tissue (Section 3.3.1), ammonite jaws and radula. Spherical concretions will generally be earlier than more flattened ones, unless the flatter upper surface was controlled by lithological change, as with a concretion in a mudrock just below a sandstone.

Impressions in primary flint and chert often display an unusual amount of information. For instance, echinoid spines extracted from white chalk are generally imperfectly preserved, however careful the preparation may be. But a latex pull of the flint impression displays the spine's delicate thorns with perfect form.

Aragonite, on the other hand, generally dissolved before the formation of flint (Fig. 3.21). Flints often enclose a hollow left by a dissolved sponge, and cracking them reveals flint meal – dusty sponge spicules, foraminifera and ostracodes. The ostracodes are partly replaced by silica. The flint meal biota is often richer and more diverse than the biota from the surrounding chalk.

Concretions of all types may be reworked intraformationally, or as pebbles and cobbles, into younger sediments where they may act as hard substrates for colonization. Intraformational reworking can be important in stratigraphy (Fig. 3.24).

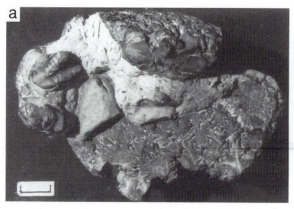
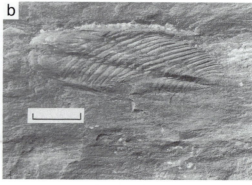

Fig. 3.24 (a) Ammonite in phosphatic concretion from Gault Clay, Kent, Lower Cretaceous; it indicates at least three stages of phosphate growth, involving repeated burial and exhumation, each indicated by differences in colour of the phosphate. (b) Deformed trilobite *Angelina sedgwicki* (Lower Ordovician, North Wales). Scale bars = 10 mm.

3.5 Dolomite replacement and silicification

Most of the modes of preservation referred to above concern mineralogical change to the actual skeleton (replacement with a void stage, or neomorphism). What has happened to the matrix has been independent of the skeleton. Wholesale replacement of the rock (skeleton and matrix) causes problems. Do not necessarily be put off by the common use of *dolomite* or *chert* in the formation name, e.g. Langland Dolomite. Examination of the rock with a hand lens may reveal partial dolomitization with scattered rhombs only. Dolomitization or silicification may be local and affect different groups differently. Echinoderm material generally resists dolomitization more than other groups. Silicification may be obvious in the field, or cryptic. Calcitic rugose corals, brachiopods and oysters are quite commonly coated by chalcedony. Wholesale silicification of the fauna in a limestone may not be easy to recognize on freshly broken surfaces. Look carefully at weathered surfaces. Trilobites, ostracodes and bryozoans have been silicified in certain limestones, but the brachiopods remain calcitic. Use a hand lens to observe acid reaction. If, for instance, silicification has been thorough then quite small (cm–dm sized) blocks may be collected for later digestion.

3.6 Breakage and deformation

Deformation of fossils is due to two processes. Compactional deformation (Table 3.4) is followed by tectonic deformation, readily distinguished by the

Table 3.4 Relative resistance to breakage for some fossils[a] and some common modern bivalves.

	Fossils[a]	Modern bivalves
Increasing resistance to breakage ↓	Jurassic ammonites[b]	*Pholas*
	Phylloceras	*Mya*
	Harpoceras	*Mactra*
	Pseudomonotis	*Mytilus*
	Inoceramus	*Pecten*
	Posidonia	*Cardium*
	Triassic ammonites	
	Ceratites	
	terebratulids	
	rhynchonellids	
	gastropods	
	crinoid ossicles	

[a] A provisional scale of natural breakage under compaction (from field observations). Adapted from Brenner and Einsele (1976).
[b] See Section 3.3.2.

structural state, cleavage, folding, etc., of the rock. Any deformation affects recognition, identification and reconstruction. This is particularly true of groups without a substantial mineralized skeleton: plants, graptolites and many ammonoids.

Fortunately, because of confining pressures, most fossils do not spread on compression; instead they crumple. Thus their lateral dimensions are virtually unchanged. Exceptions are relatively unsupported parts, e.g. apertures of graptolites and cephalopod body chambers, which fracture and push into the matrix. This is most likely to occur in very soft sediment. Uneven fracture and spread is generally obvious. Factors that lead to variation in form of the flattened fossil are decay distortion, different attitudes of

Box 3.4 A subtropical beach in southern California

Bahia la Choya (Fig. 3.25), a few kilometres north-west of Puerto Penasco (latitude 31° N), is at the northern end of the Gulf of California (a rifted basin which opened 3.5–4.5 Mya). The climate is arid and the summer surface water temperature is 30–32 °C (10–14 °C in winter). Salinities are slightly raised in shallow coastal waters and the tidal range, up to 9 m, is associated with rapid flooding. There is marked intertidal zonation of the biota. A 4 h walk across this relatively sheltered site, from the outer flats to the salt marsh, demonstrates the zonation, associated bedforms and biogenic structures.

The outer flats are composed of medium- to coarse-grained sand, with flood-oriented sand waves and superimposed smaller ripples, also ebb-oriented, or with respect to late stage drainage. Sand wave troughs are often shelly with washed-in sand dollars (*Encope*) and disarticulated bivalves. Live animals are relatively rare in this high-energy environment. Depressions excavated by rays searching for bivalves are common and locally there are patches of vertical, shell-lined burrows (*Diopatra*, Fig. 8.30). The surface trails of predatory *Natica* have no chance of being preserved as fossils, but the morphology of the trails gives a clue to the way fossil tracks, probably made by gastropods, were formed.

In contrast, the inner flats are characterized by smaller grain size and essentially small-scale, ebb-oriented, though wind-modified, ripples. There is a low topography of slightly raised areas of algal-bound, fine sand, broken by depressions of pellet-rich, soupy fine sand. The live fauna includes deep-burrowing razor shells, *Tagelus*, and the scavenging gastropod *Nassarius*. Although displaying little surface expression, the extensive burrow systems of the ghost shrimp *Callianassa* can be found by digging. If time allows, it is useful to discuss how closely the burrows resemble fossil burrows such as *Thalassinoides*. There is a concentration of shell material at a some-what variable depth, resulting from sediment-sorting activities of deposit feeders. When digging on the muddy inner flats, the boundary between oxic and anoxic sediments can be seen very close (only 1 mm)

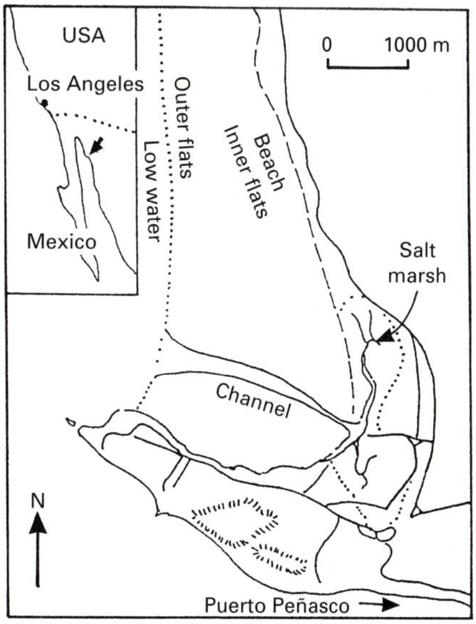

Fig. 3.25 Sketch map of Bahia la Cholla, 7 km north-west of Puerto Penasco, northern Gulf of California showing major habitats and course of channel.

to the surface, in contrast to the sandy channel sands where it is several decimetres down.

Aspects that may be usefully discussed are the variation in shell destruction in the different environments: physical and biological destruction being highest in the outer flats and channel, whereas chemical destruction, scarcely evident on the outer sands, becomes an important aspect in the preservation of salt marsh shells. In the uplifted Pleistocene deposits around the bay, only impressions or chalky remains of formerly aragonitic shells survive. An aspect that is not obvious during the traverse is the role of reworking. Some of the more robust shells have been radiocarbon dated as greater than 3000 years BP indicating extensive time averaging of the biota and loss of the more delicate shells. A more complete guide to the area can be found in Flessa (1987).

Sources: Flessa (1987), Fürsich and Flessa (1991)

---

appendages and, especially, variation in the original attitude of the organism to the bedding.

Reconstruction of compressed fossils will generally be undertaken in the laboratory. A copier that has anamorphic zoom, allowing the *x* and *y* axes to be independently enlarged or reduced is a useful technique (Rushton and Smith, 1993). The restoration technique described by Briggs and Williams (1981) is also useful. This involves photography of models and com-

paring the photographs with the compressed material. Silt, sand and coarser-grade sediment scarcely compacts unless rich in organic (plant) matter. The effects of pressure dissolution are generally obvious in limestones, but are less obvious in mouldic preservation in sandstones or siltstones.

There are sedimentological implications with compacted shells. For instance, fractured shells in a concretion may provide an indication of the burial depth at which lithification took place, if the loading threshold for shell fracture can be determined. Not enough data are yet available, however, to provide useful information for the field.

Check whether or not early dissolution took place, and do not confuse septarian cracks (tensional) with load fracture. The presence of intact shells in muddy sediment means either that the load has been low, or that the enveloping sediment became lithified relatively early. Broken shells show that breaking pressure was reached before lithification.

Tectonically deformed fossils are discussed in Section 7.7.6.

References

Ager, D. V. (1963) *Principles of paleoecology*. McGraw-Hill, New York

Allison, P. A. (1988) Soft-bodied animals in the fossil record: the role of decay in fragmentation during transport. *Geology* **14**: 979–81

Bockelie, J. F. (1994) Plant roots in core. In Donovan, S. K. (ed.) *The palaeobiology of trace fossils*. Johns Hopkins University Press, Baltimore MD, pp. 177–99

Bosence, D. W. J. (1983) Coralline algal reef frameworks. *Journal of the Geological Society, London* **140**: 365–76

Brenner, K. and Einsele, G. (1976) Schalenbruch im Experiment. *Zentralblatt für Geologie und Paläontologie Jahrgang*, 349–54

Brett, C. E. and Baird, G. C. (1986) Comparative taphonomy: a key to paleoenvironmental interpretation based on fossil preservation. *Palaios* **1**: 207–27

Briggs, D. E. G. (1995) Experimental taphonomy. *Palaios* **10**: 539–50

Briggs, D. E. G. and Wilby, P. R. (1996) The role of the calcium carbonate–calcium phosphate switch in the mineralization of soft-bodied fossils. *Journal of the Geological Society, London* **153**: 665–68

Briggs, D. E. G. and Williams, S. H. (1981) The restoration of flattened fossils. *Lethaia* **14**: 157–64

Eglington, G., Curtis, C. D., McKenzie, D. P. and Murchison, D. G. (eds) (1985) Geochemistry of buried sediments. *Philosophical Transactions of the Royal Society of London A*, Vol. 315 (A collection of papers relating to marine and freshwater diagenesis)

Flessa, K. W. (ed.) (1987) *Paleoecology and taphonomy of Recent to Pleistocene intertidal deposits, Gulf of California*. Special Publication 2, Paleontological Society (The society has a publications office at the Carnegie Museum of Natural History, Pittsburgh PA)

Fürsich, F.-T. and Flessa, K. W. (eds) (1991) Ecology, taphonomy, and paleoecology of Recent and Pleistocene molluscan faunas of Bahia la Choya, northern Gulf of California. *Zitteliana* **18**: 1–180

Hutton, A. C. (1986) Classification of Australian Oil Shales. *Energy Exploration and Exploitation* **4**: 81–93

Martill, D. M. (1988) Preservation of fish in the Cretaceous Santona Formation of Brazil. *Palaeontology* **31**: 1–18

Rex, G. (1983) The compression suite of preservation of Carboniferous lepidodendroid leaves. *Review of Palaeobotany and Palynology* **39**: 65–85

Rex, G. (1985) A laboratory flume investigation of the formation of fossil stem fills. *Sedimentology* **32**: 245–55

Rex, G. and Chaloner, W. G. (1983) The experimental formation of plant compression fossils. *Palaeontology* **26**: 231–52

Rushton, A. and Smith, M. (1993) Retrodeformation of fossils – a simple technique. *Palaeontology* **36**: 927–30

Seilacher, A. (1971) Preservational history of ceratite shells. *Palaeontology* **14**: 16–21

Seilacher, A. (1976) Preservational history of compressed Jurassic ammonites from southern Germany. *Neues Jahrbuch für Geologie und Paläontologie Abhandlungen* **152**: 307–56

Smith, A. (1984) *Echinoid palaeobiology*. George Allen & Unwin, London

Speyer, S. E. and Brett, C. E. (1988) Taphofacies models for epeiric sea environments. *Palaeogeography, Palaeoclimatology, Palaeoecology* **63**: 225–62

Spicer, R. A. (1991) Plant taphonomic processes. In Allison, P. A. and Briggs, D. E. G. (eds) *Taphonomy: releasing the data locked in the fossil record*. Plenum, New York, pp. 71–113

Stach, E. (ed.) (1982) *Stach's textbook of coal petrology*. Gebrüder Borntraeger, Berlin

Taylor, P. D. (1990) Preservation of soft-bodied and other organisms by bioimmuration – a review. *Palaeontology* **33**: 1–17

Taylor, P. D. and Todd, J. A. (1990) Sandwiched fossils. *Geology Today* **6**: 151–54

Wilby, P. R. and Briggs, D. E. G. (1997) Taxonomic trends in the resolution of detail preserved in fossil phosphatised soft tissues. *Geobios* **20**: 493–502

Wilby, P. R., Briggs, D. E. G., Bernier, P. and Gaillard, C. (1996) Role of microbial mats in the fossilization of soft tissues. *Geology* **24**: 787–90

Further reading

Most palaeontological and palaeobotanical texts (see Chapter 5) have sections on taphonomy and fossil preservation. Some useful, but detailed and specific, references to the state of the art are included below. And two journals have special issues on taphonomy: *Palaios*, Vol. 1, no. 3 (1986); *Palaeogeography, Palaeoclimatology, Palaeoecology*, Vol. 63 (1988).

Allison, P. A. and Briggs, D. E. G. (eds) (1991) *Taphonomy: releasing the data locked in the fossil record*. Plenum, New York

Donovan, S. K. (ed.) (1991) *The processes of fossilization*. Belhaven, London

CHAPTER 4

Fossil identification

Identification is a necessary chore but it is lineage that matters!

This chapter describes how to set about the identification of body and trace fossils, and to recognize pseudofossils. The significance of misidentification is also discussed.

4.1 Body fossils

Is the mode of preservation the same? This is the first question to ask when making comparisons with illustrations or with specimens in museums, etc. If not, then allowance has to be made. This is a more complex topic than it may seem, as can be appreciated from Chapter 3. There are also many instances of different names having been given, and sometimes still applied, to different preservation states of the same taxon. Strictly speaking, these instances are unacceptable under international rules. For example, in fossil plants, *Arthropitys* is given to the anatomical mode and *Calamites* to compressions and pith casts of Carboniferous sphenopsids (horsetails). *Knorria* is applied to decorticated impressions of *Lepidodendron*. Also, where different parts of a fossil plant cannot yet be confidently associated because, for instance, leaves, seeds and stems have been separated (Section 3.2), it is normal to name each part differently; a procedure called parataxonomy. Parataxonomy is also applied to the dispersed skeletal elements of conodonts, but fortunately it is not generally used for other vertebrate fossils.

When making an identification, it pays to have a good appreciation of the material. Make sketches (e.g. plan view, side view) at a suitable magnification, labelling features that seem likely to be diagnostic. If the identification book has a key or diagnoses, peruse them to ascertain what features are diagnostic. The

morphological features used to establish fossil taxa (taxobases) are too numerous to discuss here, and keys for fossil taxa are not practicable. Taxobases range from the relatively simple, e.g. the number of rows of plates in the columns of echinoids, to the subtleties of curvature along the suture line in a Carboniferous trilobite. Do not be surprised to find diagnostic characters missing from the material to hand because of poor or incomplete preservation. For example, besides general shape and size, a brachiopod diagnosis will refer to internal features such as dentition and muscle attachment areas. Without these features, identification may be possible only to genus or family level. Most handbooks of fossils give only the name, age and possibly other names that have been applied to the fossil in the past, in addition to the illustration. And handbooks usually carry only a selection of the more common fossils. Local guides, map explanations and accompanying memoirs often have faunal and floral lists for specified localities, and it may be helpful to scan them.

The names of living and fossil organisms are formed on classical lines governed by codes (Greuter *et al.*, 1994; ICZN, 1999). It is sensible to know something about these codes. A useful guide to the construction and meaning of names is in Jaeger (1955). Identification can be to the genus, species or to a higher rank (e.g. family). A species name, e.g. *Agenus beta* Jones, 1915, has three parts: the generic name, the specific name and author(s), generally with date of publication. (The specific part may be multiple, e.g. with addition of a subspecific name.) If the specimen can only be identified to generic level, then use *Agenus* sp. A still lower level of identification would be to an orthid brachiopod, one of a handful of groups into which the phylum Brachiopoda is divided. It is useful also to check the validity of the illustration being used.

Does the caption mention whether the illustration is the *holotype* or the *lectotype*? The holotype is the single specimen to which the name validity belongs as stated by the author; the *lectotype* is the type specimen chosen later on from a number of *syntypes*. Among the other possibilities, *topotype* is the most useful; the topotype is material from the same locality as the holotype or the lectotype.

Line illustrations are often reconstructions. They may look nice but are subject to error by the artist. Do not rely on the description alone as it may contain terms that are ambiguous or of uncertain meaning. Even type material may have been altered; for example, a productid brachiopod in a national collection had had its endospines inadvertently removed. It is also useful to remind ourselves that a fossil species isn't like a living biospecies. If a biospecies can be loosely defined as a population of actually (or potentially) interbreeding animals or plants, then a fossil species is merely an assemblage of morphologically similar fossil remains! Fossil genera are lineages.

When an illustration and description that agrees with the material to hand has been found, ask the question, How close is the agreement? Relatively few descriptions will be accompanied by illustrations covering the complete range of variation shown by a taxon, or even a full (statistical) description of the variation. Check the account to see what variation is referred to. For taxa that are known only by a few specimens, it is likely that new material, especially if acquired from a new locality, will show some differences. Material of common taxa from well-known localities, is likely to fall within the known range of variation. If the material differs from the published accounts, then it can represent one of several things:

- An extension in the variation of the taxon.
- A subspecies characteristic of a geographical area or geological horizon.
- A taxon that bears only general comparison with a described species, and may be referred to as *Agenus* cf. *beta*.
- A close new taxon, which may be temporarily designated as *Agenus* n. sp. aff. beta.
- A greater degree of uncertainty may be indicated as *Agenus beta*? or, where specific identification is impossible, but the genus appears correct, as *Agenus* sp.
- A new taxon *Agenus* sp. nov., and the responsibility to ensure that the material is properly curated and that the discovery is made known by publication!

A further problem in identification concerns material from immature or gerontic individuals. For fossils that display continuous growth (brachiopods and bivalves), this is generally readily recognizable; but careful consideration and allowance is required with fossils of groups that grow by moulting and/or addition of new parts, or by modification of parts. Identification is not as objective an operation as might be supposed!

4.1.1 Sources for identification

Start with local guides and local museums, and local and national illustrated handbooks. Refer to range charts as a guide, but do not expect them to be infallible. Use the bibliographies as an introduction to more specific monographs and papers, many of which may be available only from reference libraries in larger cities, universities, etc.

Illustrations and descriptions are published in a wide range of journals and occasionally in books; they generally deal with either a specific area or a specific group of fossils. Here are some of the journals: *Alcheringa, Acta Palaeontologica Polonica, Beringeria, Fossils & Strata, Journal of Paleontology, Journal of the Paleontological Society of India, Monographs of the Palaeontographical Society, Paleontographica Americana, Paleontographica Canadiana, Palaeontology, Paläontologisches Zeitschrift, Senckenbergiana Lethaea, Special Papers in Palaeontology, University of Kansas Paleontological Contributions*.

Some of these journals have cumulative indexes, which will assist in locating papers. There are other major compilations such as the ongoing *Treatise on Invertebrate Paleontology* (Geological Society of America), where each volume deals with a phylum or class. The *Treatise* is concerned almost entirely with taxa at genus level.

These steps in identification do not mention getting an identification from a local expert, or sending or taking material to experts in research institutions and national museums, etc. If you have taken some trouble yourself, have localized the material, and gone some way to understanding it, then the 'experts' are nearly always glad to help!

Most references to the identification of fossils give little or no ecological information. Palaeoecological interpretation has to be tackled in a systematic way (Chapter 5), working from the individual fossil or species (autecology) to the assemblages and communities (synecology). For fossils with living representatives (even at family level or higher rank), it always pays to assess the extensive literature, videos, etc., on the living biota.

4.1.2 Misidentification

Stratigraphic error The level of a stratigraphic error depends on the stratigraphic refinement required for the job in hand, and the significance that others might place on your identification for regional correlation or tectonic structure and history. Identification to the zonal level (Section 7.4) may require considerable skill. Even identification to system level can cause problems, e.g. the much quoted howler of the European graduate who identified an Australian Cretaceous ammonoid with a 'ceratitic' suture line as Triassic in age!

Ecological error Misidentification that implies an *ecological* error. A likely case is identification to a taxon from a different geographical province or climatic zone, which might have serious consequences for palaeogeographic reconstruction.

Out-of-date naming Giving a name that is 'out of date' because of revision is generally not a problem. If the source for the identification is quoted in your report then an expert will understand the error.

Unacceptable renaming Forming a new name where a valid name already exists is unacceptable and won't be accepted by a refereed journal.

4.1.3 Ancestors and descendents

The classification of fossils is concerned with interrelationships. There are several methods of producing a 'natural' classification, and these are covered in the texts listed. It is likely, at least at an initial stage, that *morphological resemblance* between one taxon and another will form the basis. More sophisticated approaches involve using *quantitative* methods. There is also the *cladistic* approach, which emphasizes shared features (Smith, 1994).

4.1.4 Some commonly misidentified groups

Ammonoids Gastropods Serpulids	Occasionally there are septa in Carboniferous vermetid-like gastropods Gastropod shells are generally thicker Coiling in serpulids is more or less open ± flanges, and growth is often slightly irregular
Scaphopods Serpulids	Tusk-like serpulids in the Cenozoic can be distinguished from scaphopods by calcitic shell and closed apex
Brachiopods	Bilateral symmetry, inequivalve (except for rare exceptions) Calcitic shell can be punctate, pseudopunctate Spines, if present, are delicate
Bivalvia	Equivalve and asymmetrical, except oysters and scallops Oysters are inequivalve Scallops have almost bilateral symmetry and a single (posterior) muscle off-centre
Barnacle plate	Typically diamond-shaped with bevelled edge
Massive bryozoans	Orderly thecae
Sponges	Irregular mesh
Bone	Fine mesh, phosphatic
Echinoderm	Very fine, calcitic mesh of stereom
Stromatolites Oncolites	Irregular lamination, no vertical elements
Stromatoporoids	Fine brickwork-like skeleton ± mamelons
Solenopores (rhodolites, red algae)	Still finer, and more regular, 'conifer wood in cross-section', cellular structure
Rudists Corals	Small rudists may look like solitary corals, but they lack septa
Roots Rhizoliths	Downward bifurcation, with decreasing width Roots often associated with carbonaceous film and slickensiding Pneumatophores (air-seeking structures) grow upwards; they are present in some swamp-dwelling plants, e.g. mangroves
Burrows	Downward bifurcation with uniform width; burrows with menisci or spreite, faecal pellets, variable direction
Modern bone	Greasy and gives off pungent odour when a lighted match is applied
Fossil bone	Much heavier, cells ± infilled
Fern leaf	Symmetrical midrib
Cockroach wing	Curved midrib
Fern leaf Seedfern leaf	Not possible to distinguish without attached seed or the presence of sporangia; both conditions are rare
Angiosperm leaf	Blind-ending veinlets
Gymnosperms with angiosperm-like leaf	Veins form complete network
Charcoal (fusinite)	Black streak on paper
Bog oak, jet, vitrinite	No streak

4.2   Trace fossils

Trace fossils are given binomial names (in Latin) in the same way as body fossils, but this is more a matter of convenience and history than of taxonomic consistency. There are probably at present only about 300 ichnogenera (igen.) and not that many more ichnospecies (isp.). Many ichnogenera are still poorly understood. There are no generally used higher ranks of classification. The *Treatise on Invertebrate Paleontology* (Part W) lists ichnogenera alphabetically, as do most larger compilations and monographs. These can be found in the same journals and other publications listed in Section 4.1.1, and also in the journal *Ichnos*. But several authors group ichnotaxa of like behaviour (e.g. resting traces, Cubichnia), or similar morphologies (e.g. radial traces), or by attributing to a particular group of (supposed) trace maker. Kuhn (1963) lists vertebrate trace fossils (illustrated by line drawing).

The naming of animal trace fossils is based on the *International Code of Zoological Nomenclature* (ICZN, 1999), but many trace fossils were considered to be the remains of seaweeds till quite late in the last century, and others regarded as 'Problematica'. This connection can be seen in some of the names that have been given, e.g. those ending in -phycus (*Palaeophycus*), or *Chondrites*, for a fossil *Chondrus* (a marine alga). Several articles in the code refer to trace fossils under the heading 'work' of an animal.

Many trace fossils are large and/or ramifying sedimentary structures and are rarely exposed in full. It may be necessary to do much field preparation. As likely as not, much of the trace may be missing through weathering, or just because of being reworked by another tracemaker. Many identifications are made on only part of the structure. For instance, *Ophiomorpha nodosa* may be applied to a short length of a pellet-lined gallery. But strictly, the name applies to the network or boxwork of shafts and galleries with a specific mode of branching and nodes. This is where description of the material to hand is important. An 'incomplete' burrow system or unusually narrow burrow may have environmental significance (Section 8.5.8). Poor preservation may mean giving a name that could be misleading and the term *taphoseries* has been applied for such situations (Fig. 8.1). *Compound* trace fossils refer to one type of trace passing into another. As an example, if a boxwork of *Ophiomorpha nodosa* shows a section that is unlined, this section could be assigned to *Thalassinoides*. But it is simpler to note the local variation. Trace fossils are described as *composite* when secondary burrows are present, e.g. *Planolites* within the fill of *Thalassinoides*.

The features (ichnotaxobases) used to name a trace fossil are much less clear-cut than the taxobases for body fossils. Unfortunately, most are without any proper foundation in respect of the process by which the trace was formed and, with some notable exceptions, few experiments have been carried out to determine exactly how different traces were formed: by the animal in advancing through or over the substrate surface and how the sediment responded to the intrusion. There is much scope for further work, but the problems are considerable.

Most ichnotaxa are distinguished on the basis of general form, type of branching, the nature and structure of the burrow fill and suchlike features. Fürsich (1974) suggested extending this morphological analysis to an assessment of which features were considered to be of significance and thus of ichnogeneric status, and features that were 'accessory' and of ichnospecific status. But much subjectivity is involved with this approach. Unfortunately, many ichnotaxa were given names without the authors asking the question, What was the form of the burrow occupied by the producer? And then asking, How was the burrow extended? In addition, the question has to be asked, How did the nature of the substrate influence the burrow morphology, or foot penetration? Fig. 4.1 illustrates some of the basic concepts. It is proposed that the *primary* ichnotaxobase is the *morphology of the original burrow, burrow segment or trace* and that the *secondary* ichnotaxobase is the *manner in which the burrow is displaced and/or extended* (Goldring et al., 1997). However, the paradox is that it is often useful to use names that strictly refer to specific taphonomic effects just because they carry important sedimentological and palaeoecological information (Fig. 4.2). One solution is to use a dual naming. For instance, with reference to Fig. 4.2:

Ichnogenus *Scolicia* de Quatrefagas, 1849
Toponomic expression *Subphyllochorda*
        Gözinger & Becker, 1932

Or if the sediment has split so as to reveal only one aspect of a complex burrow, material is limited and mechanical preparation (e.g. needle) does not provide more evidence, then it is useful to refer to the likely ichnotaxon and the ichnotaxon associated with the particular preservational style. For instance, if only a bedding-parallel section is evident, it might be referred to as

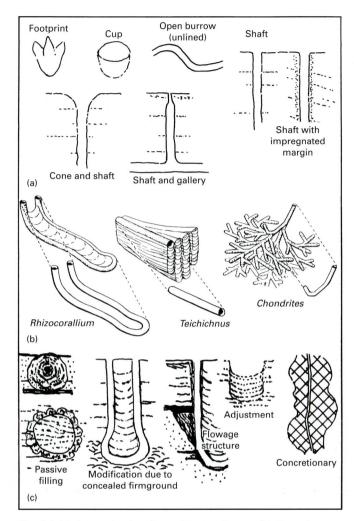

Fig. 4.1 (a) Some simple primary ichnotaxobases. (b) Three burrows and their displacement or extensions (secondary ichnotaxobases). (c) Some common features of trace fossils which may not serve as ichnotaxobases. The diagrams in part (b) are adapted from Seilacher (1957), who drew them for a different purpose.

Ichnogenus *Scolicia*? de Quatrefagas, 1849
Taphonomic expression *Laminites* Ghent & Henderson, 1966

The other problem in naming trace fossils is in the limited exposure seen on slabbed core or rock face. Fig. 8.18 shows some examples of ichnotaxa in core expression. For other ichnogenera, attempt a block diagram and then explode it. Allow for compaction and diagenesis. Many ichnospecies are distinguished, probably wrongly, by size. But the subtle variations of the scratch marks made by the appendages of stratigraphically useful *Cruziana* (most are attribut-

able to trilobites) are usefully assigned to ichnospecies (Fig. 8.12).

Naming arthropod tracks and vertebrate footprints

1 The lightly impressed but often sharp elements of arthropod appendages may penetrate several laminae to form an *undertrack fallout* (Fig. 8.1). Different appendages will have penetrated to different depths. The name for the trackway should refer to the *highest* level of organization that can be recognized. Less definite tracks at other levels

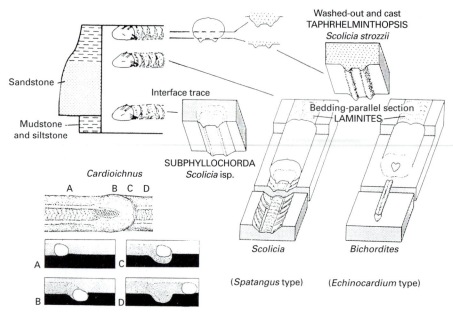

Fig. 4.2 Scolicia group trace fossils attributed to burrowing echinoids, with toponomic and taphonomic expression. Names in capitals are those commonly used and indicating toponomy and taphonomy; names in italics are used when the preservational style is acknowledged. The lower left diagram is a model for the formation of *Cardioichnus* (B), echinoid 'resting' trace, linked to *Scolicia* (A,D), a compound trace fossil.

are referred to as 'an undertrack of ichnogenus B'.

2 The footprints of heavy vertebrates may strongly deform the sediment. The name for the print should be based on specimens with greatest relief. Ideally this will be the natural footprint (or artificial cast made from this). Natural casts will nearly always show evidence of erosion, etc. *Transmitted* footcasts (Fig. 8.1) will also display less information with low relief.

4.2.1 Misidentification

Trace fossils are not the remains of animals or plants, only sedimentary structures caused by activity. It is illogical to treat trace fossils exactly as if they were the remains of animals or plants. There are five categories of error in the identification of trace fossils:

1 The name reflects only minor differences in behaviour. For instance, reference to *Planolites* (a simple, backfilled burrow) instead of *Helminthopsis* (a backfilled burrow of similar form but meandering course). However, the latter is only known from marine facies.

2 The identification relates to a quite different behavioural style, and hence ecological interpretation.

For instance, the spreite of *Diplocraterion* (the work of a tube-dwelling suspension feeder) can be mistaken for *Teichichnus* (a retrusive deposit feeder).

3 The identification implies an incorrect palaeoenvironment. For instance, *Ophiomorpha* has been wrongly applied a number of times to the backfilled burrow *Beaconites*. *Ophiomorpha* always indicates a marine association, whereas *Beaconites* is typical of non-marine environments.

4 Some trace fossils may resemble body fossils, e.g. serpulids.

5 Some trace fossils resemble sedimentary structures (Fig. 4.4).

4.3   Pseudofossils

Two-thirds of Earth's history is represented by Precambrian rocks, and it is only towards the top of the Precambrian that evidence of larger organisms can be recognized (Section 8.6). There is, though, abundant evidence of simple plants and trace fossils (associated with microbial mats) extending far back into the Precambrian. The temptation to be the first to find the oldest metazoan is strong, but it is wise to be aware of possible pitfalls (Fig. 4.3), and the criteria to use on

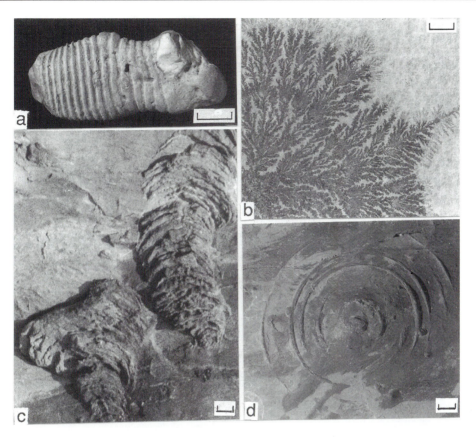

Fig. 4.3 (a) Banded flint suggesting annelid segmentation (Upper Cretaceous). (b) Dendritic markings suggesting moss, due to manganese oxide on joint plane. (c) Gas escape structures formed in cross-bedded sandstone at the margin of a scour or channel fill structure which had a large log at the base; the structure superficially resembles the trace fossil *Beaconites* (Clockhouse Brickworks, southern England, Weald Clay, Lower Cretaceous). (d) Tool mark (scratch mark) on sole of very fine-grained sandstone cut into mud, probably formed by plant swinging in a fluctuating current (lacustrine facies, Warnham Brickworks, Sussex, England, Weald Clay, Lower Cretaceous). Scale bars = 10 mm.

structures of uncertain origin. Although this applies particularly to Precambrian sediments, it is also relevant to Phanerozoic sediments, since the same sorts of problem occur there. There are three categories of false lead.

Activities of modern animals For example, burrows in Precambrian sediments, which have subsequently been found to be the work of modern termites probing deep to get to the water table; or burrows, also in Precambrian sediments, which were later recognized as due to modern annelids, penetrating into joints and slightly dissolving the calcareous sediment. The clue is to look carefully at the bedding relationships (Section 2.3).

Structural or sedimentological errors For example, in a belt of Precambrian quartzites a unit

of downfaulted Tertiary sandstones with burrows was initially unrecognized as such. In Siberia, in late Precambrian dolomites, a layer of fossils was unrecognized as the infill of a sedimentary (Neptunian) dyke system from the Lower Cambrian.

Pseudofossils Pseudofossils are rock structures that resemble fossils, that look like plants or animals or parts of them, or look like the activities of animals. It is important to pose the right questions to decide whether a structure is a fossil or a pseudofossil. First consider taphonomic aspects and the organization of sedimentary structures (bedding, lamination). There is now a long list of structures that resemble animals, animal activity or plants to which Linnaean names were applied, but which have now been shown to be of inorganic origin (*Treatise*, Part W).

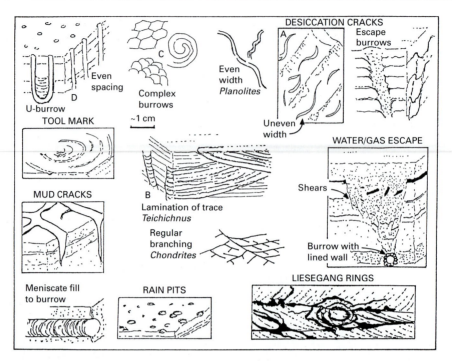

Fig. 4.4 Field evidence that might be used to distinguish between trace fossils (unboxed) and some common Phanerozoic primary sedimentary structures (boxed): (A) desiccation cracks in mud ripple troughs; (B) lamination formed by upward migration of the *Teichichnus* animal, which may sometimes resemble small-scale cross-lamination; (C) complex burrows (Figs 8.4 and 8.13); (D) U-burrow (Fig. 8.19) and vertical single burrows.

Eliminating false leads

It is generally easier to try to find a sedimentological solution than to obtain proof of biological origin. But, in the late Proterozoic especially, one must anticipate that sediment deposition did not take place under the same hydraulic conditions as can be observed today. This is because microbial binding of sand grains (as well as muds) was widespread, and led to sand-flake intraformational conglomerates and even rolled-up sand sheets as well as other peculiar structures. Referring to Fig. 4.4, gas or water escape structures often look like burrows, but an upward cone with sheared walls is not the structure produced by an animal moving upwards. If the sediment is loose, confirmation of an inorganic origin can often be found by probing to the base of the cone to a ruptured burrow. Sinuous, tapering desiccation cracks in ripple troughs (the so-called Manchuriophycus), or polygonal V-shaped desiccation cracks, likewise differ from the even width of a burrow. The irregularity in size of rain pits distinguishes them from the more uniform width of burrow openings. Here try to locate the burrows extending downwards normal to bedding. Certain tool marks (Section 6.6) are probably due to current action on

attached weed (Fig. 4.3). Liesegang rings or bands, secondary structures caused by rhythmic precipitation (generally of iron oxide) in a fluid-saturated rock, often resemble burrows. Primary sedimentary structures, such as cross-lamination passing across the rings, demonstrate their secondary origin. Similarly, primary lamination passing across concretions is proof of their stratinomical relationship. But concretions which have a body-fossil-like shape may nevertheless encase true fossils (Section 3.4), or have formed along burrow fills. Dendritic markings (Fig. 4.3(b)) caused by deposition of a manganese film along bedding or joint surfaces are common pseudofossils. Diagenetic structures usually give themselves away because of the mineralogy (pyrite, flint).

One should use one's knowledge of evolution with confidence. For example, apparent fragments of a eurypterid in positively identified Precambrian rocks, cannot derive from a eurypterid. Closer inspection would soon reveal the absence of scales on the supposed carapace plates, which turn out to be intraformational mud chips. For macrostructures, highly evolved structures and growth rings indicating ontogeny provide evidence of biological origin. Simple symmetry and abundance are not acceptable criteria. The category *dubiofossils* has been proposed to cover such cases of doubtful origin!

There are also some true organic structures that often look inorganic, such as stromatolites and oncolites (Section 3.2.3). It is often exceedingly difficult to go further than indicate a strong similarity with modern examples. In most cases interpretation is dependent on slabbing.

References

Fürsich, F. T. (1974) On *Diplocraterion* Torell 1870 and the significance of morphological features in vertical spreiten-bearing, U-shaped trace fossils. *Journal of Paleontology* **48**: 952–62

Goldring, R., Pollard, J. E. and Taylor, A. (1997) Naming trace fossils. *Geological Magazine* **134**: 265–68

Greuter, W. *et al.* (1994) *International code of botanical nomenclature.* Koeltz, Königstein (ISBN 3-87429-367-X)

International Commission on Zoological Nomenclature (1999) *International Code of Zoological Nomenclature (4th edition)*, International Trust for Zoological Nomenclature, London (ISBN 085301 006 4)

Jaegar, E. C. (1955) *A source book of biological names and terms.* Charles C. Thomas, Springfield IL

Kuhn, O. (1963) *Fossilium Catalogus 1: Animalia, part 101 Ichnia Tetrapodorum.* W. Junk, The Hague (Text and outline figures only)

Seilacher, A. (1957) An-aktualistiches Wattenmeer? *Paläontologischs Zeitschrift* **31**: 198–206

Smith, A. B. (1994) *Systematics and the fossil record.* Blackwell Scientific, Oxford

Further reading

Publications of the Natural History Museum, London

British Palaeozoic Fossils
British Mesozoic Fossils
British Caenozoic Fossils

Field Guides published by the Palaeontological Association, London

Collinson, M. E. (1983) *Fossil plants of the London Clay* (Eocene)

Smith, A. B. (1987) *Fossils of the Chalk* (Upper Cretaceous)

Hollingworth, N. and Pettigrew, T. (1988) *Zechstein reef fossils and their palaeoecology* (Permian)

Martill, D. M. and Hudson, J. D. (1991) *Fossils of the Oxford Clay* (Upper Jurassic)

Martill, D. M. (1993) *Fossils of the Santana and Crato Formations, Brazil* (Upper Cretaceous)

Cleal, C. J. and Thomas, B. A. (1994) *Plant fossils of the British Coal Measures*

Harper, D. A. T. and Owen, A. W. (1996) *Fossils of the Upper Ordovician*

Other publications

Gillespie, W. H., Latimer, I. S. and Clendening, J. A. (1966) *Plant fossils of West Virginia.* West Virginia Geological and Economic Survey, Charleston, IO

Hill, D., Playford, G. and Wood, J. T. (1964–1973) *Illustrations of fossil faunas and floras of Queensland.* Queensland Palaeontographical Society, Brisbane

Murray, J. W. (ed.) (1985) *Atlas of invertebrate macrofossils.* Longman, Harlow (With several keys, but without archaeocyathids or trace fossils; the English edition is out of print)

Murray, J. W. (ed.) (1985) *Wirbellose Makrofossilien.* Ferdinand Enke Verlag, Germany (ISBN 0049 711 2030)

Pemberton, S. G. (1999) *Ichnological atlas of clastic environments: applications to exploration and production geology.* American Association of Petroleum Geologists, Tulsa OK

Treatise on invertebrate paleontology

An ongoing series of volumes covering most groups of invertebrates, including trace fossils and problematica, published by the Geological Society of America and (for earlier volumes) the University of Kansas. Besides systematic information, many volumes contain detailed descriptions of the biology, physiology and ecology of the modern representatives.

CHAPTER 5

Body fossils for the palaeontologist and palaeoecologist

Every fossil was once a living organism.

Palaeontology aims to determine the origin and evolution of the biosphere. This means recognizing and describing each taxon, knowing its origins, geographical distribution, relationships with other taxa and its demise. It is irrelevant that much of the story is lost because of the incompleteness of the fossil record. The quest for the patterns of evolution is there. The processes of evolution are best left to the biologists.

Every year sees significant new finds made by geologists at all stages of training and levels of experience. The sorts of information that can be unearthed include recognition of new taxa; better information on the morphology of poorly known taxa; information on geographical distribution; a new mode of preservation that allows better understanding of biological affinities; specimens showing new information on physiological functions; trace fossils (Chapter 8) that show information about behaviour; and more precise information on stratigraphic occurrence.

The aim of this chapter is to encourage the search for new evidence, as well as to indicate how to undertake field investigations, make a census, or palaeoecological analysis. It is useful to separate the ecological and systematic aspects of fossil material. The ecological use of body and trace fossils is of wider interest to geologists, and tries to answer the questions (Table 5.1) how, where and with what? For the systematist concerned primarily with morphology and evolution, the questions are what, when, whence, whither? In the field one is concerned with recognizing just what the material is, from phylum level right down to the species. This means having a pocket field-identification volume (Chapter 4).

Many consider that palaeoecology can only contribute to palaeoenvironmental interpretation, and that it is not really feasible to attempt to determine what the productivity and energetics might have been for

Table 5.1 Basic questions to be asked for each fossil.

To *what* group does it belong?	Systematic assignment
When was it a living organism (or part of, or associated with a living organism)?	Time aspect
Whence did it evolve?	Ancestry
Whither did it evolve?	Descendants
Who (if anyone) has named and described it?	Taxonomy
How did it live?	Palaeophysiology
Where did it live?	Habitat
With what was it associated?	Community analysis

ancient sediments. But there are a number of fossiliferous sites where the preservation is so good that reasonable figures could be obtained (Powell and Stanton, 1985). Even shell beds (Chapter 6) that are clearly allochthonous may, in certain cases, be regarded as time-averaged and representative of the environment in the long term.

5.1 Palaeontological and palaeophysiological analysis

Palaeontological analysis of morphology, ontogeny, functional morphology and evolution requires particular types of evidence: completeness of individuals, perfection of detail and often original shell structure. The classification of fossiliferous sediments (Chapter 1) gives an indication of the most rewarding types of sediment; see also the section on fossil ores (Section 5.5).

Try to answer the question, What mode of preservation will be most useful in providing the information

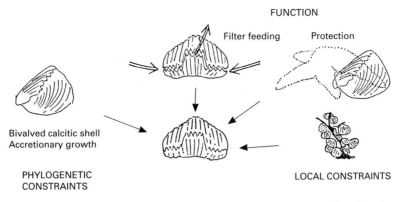

Fig. 5.1 Major factors influencing skeletal morphology, as illustrated by *Rhynchonella*, a calcitic bivalve shell with accretionary growth.

required? In every case one is collecting for subsequent preparation and examination. Material may be limited, as with small pockets in a scour-fill structure, or just the exceptional 'lucky' occurrence. It is important to make and record observations about the site before extracting any fossils. Sketches and measurements of orientations may prove useful later. A graphic log and photographs may also be useful (Section 2.4). Evolutionary studies require sufficient numbers of particular taxa, and of different growth stages. In evolutionary studies it is important to get the stratigraphy absolutely right. Ask yourself, Is there a strike fault that repeats the stratigraphy, or is there a stratigraphic break in the section? Record on log sheets or in a notebook the exact location of samples collected. If a change in the biota is spotted through a sucession that might be attributable to evolutionary processes, then it is always worthwhile considering the possible evolutionary pattern (punctuated equilibrium or phyletic gradualism) and the process that was involved (isolating mechanisms or change in timing of ontogenetic events – heterochrony).

In palaeophysiological analysis one is trying to determine just how a fossil 'lived', how it once fed and on what, how it respired, grew, reproduced, moved and protected itself. Methods of feeding used by modern animals are shown in Appendix C. Skeletal morphology is influenced by several factors (Fig. 5.1). A fuller discussion is given in Seilacher (1991).

- Phylogenetic factors, e.g. the inherited calcitic shell of the brachiopod and its accretionary growth.
- A 'genetic pull' between two or more functions during evolution, e.g. feeding and protection.
- The design of an efficient mechanism for a particular function, e.g. filter feeding.

A filter feeder needs (a) a 'sieve', provided by the slender gape of valves and the lophophore within; (b) a current, produced by cilia on the lophophore; and (c) separation of inhalant and exhalant currents, by the fold and sinus.

In addition, local factors may influence development of the perfect design, e.g. crowding. Here are the 'specifications' for some physiological processes:

- *Grazing limited food resources*: guidance system to cover maximum area with minimum effort.
- *Protection for shells*: envelope that is strong and rigid, that completely encloses soft parts and does not interfere with other vital functions.
- *Reproductive mechanisms*: release of gametes so that adequate fertilization takes place under conditions for adequate dispersal and minimal loss.
- *Predation*: most predators require an effective means to locate, capture, kill and ingest prey.
- *Respiration*: external system (gills, lungs, tube feet) for exchanging respiratory gases; circulation system; provision to prevent recirculation or choking.

To satisfy oneself about the function of any structure, consider the specification for the function in mind, and how well the fossil structure meets this. More details about functional morphology may be found in texts, e.g. Boardman *et al.* (1987) and Clarkson (1998); see the further reading. Perhaps the most sophisticated example of revitalization is Paul and Bockelie's (1983) analysis of a cystoid echinoderm. Certain trace fossils can provide information about morphology, e.g. trilobite tracks and trilobite dactyli, vertebrate footprints and casts, some bivalve and gastropod traces.

Sexual dimorphism and polymorphism are now recognized in many groups of fossils. For most groups they can be substantiated only by demonstrating the close similarity between the early stages of development and identical stratigraphic range, and completing a detailed facies analysis. Similar changes in size of two 'species' through a succession may be a clue to their sexual relationship, but check that similar changes do not occur in other taxa (i.e. that the change is ecological). Dimorphism and polymorphism also result from the alternation of sexual and asexual reproduction, but this will be evident in the field only in some larger foraminifera. Mature, 'adult' ammonites display uncoiling of the umbilical seam, modification of ribbing near to aperture (e.g. lappet), and close spacing and modification of the last septal sutures. If stunting (arrested or hindered growth) is suspected in a fauna, the only way to prove this is by (a) growth analysis, plotting growth line spacing against increments in size in the suspect suite and in a normal suite; (b) identification of mature individuals of small size. Likely causes of stunting are low salinity (Section 5.3.5), oxygen deficiency (Section 5.3.6) and food deficiency. Much has been done on tree ring analysis as an indicator of ancient climate; growth banding is present in many invertebrate groups and it has been established to have a definite periodicity in corals (Figs 1.2 and 1.4) and several groups of the Bivalvia. Little palaeontological work has been done in this area as yet, but there is potential for the determination of life spans and carbonate production, to obtain clues about sedimentation rate.

5.2   Palaeoecological analysis

An ability to identify organisms accurately is an essential prerequisite for most ecological studies – and for palaeoecology. A full palaeoecological analysis of any sedimentary unit is a demanding task. Much of the information needed can be gleaned by careful observation without removing fossils or sediment. After all, what is required is to determine the relationship between the fossil and the associated sediment, and the relationships between the various taxa. Observations and deductions made on each taxon (autecology) or significant morphological features will be synthesized later (synecology). But palaeoecology is concerned with more than just fossils. If ecology is the study of the interactions of living organisms with one another and with the physical environment, then palaeoecology concerns the same interactions that existed in ancient environments. The problem is to identify what the

ancient environments were. All there is to go on are the sediments (facies) and the fossils. Thus, to be successful with palaeoecology, a good basis in sedimentology as well as in biology and chemistry is required – an integrated approach.

The first question to ask is whether or not the fossil is in the position in which it lived, i.e. is it autochthonous? (For a planktonic or nektonic organism the question is, How far has it drifted and has it been reworked?) If not, has it been disturbed (parautochthonous) or winnowed, or has it been substantially shifted (allochthonous)? In most cases the answer is clear, but it is worth calling to mind the complexities of a modern tropical reef; for instance, one with a substantial coral framework, but with patches of shell and coral sand that have formed by hydraulic action across and through the reef. A contrasting situation is the shell bed formed by a storm that has drawn numbers of starfish and crab scavengers to feed on the dead and dying. Or consider the situation where the top of a storm-generated shell bed becomes a substrate for colonization. Any accumulation may grow by successive destructive and constructive events, a situation analagous to a human settlement (a tell).

Many organisms are able to survive displacement and skeletal damage during transport. This is particularly true of many calcareous algae, where fragments of the plants may continue to grow in their new habitat. The effect of transport with such algae is a bit like pruning a hedge: it results in an abundance of new 'shoots' to produce relatively dense rhodoliths (red algae) or oncolites (blue-green algae).

Allochthonous shell accumulations (Chapter 6) have been extensively used in palaeoecology. This is largely justified by arguing that, for a particular facies, the shell beds are found to be compositionally similar. But to establish this requires careful analysis of the hydraulic processes involved, and analysis and comparison with the composition of adjacent facies, e.g. how the shelly composition of mudstones between storm layers compares with the shelly composition of the storm layer.

5.2.1   Autochthony, parautochthony and allochthony

Criteria to determine autochthony and parautochthony are listed below. When examining a fossil occurrence, it is important at the outset to look carefully at the hydraulic and sedimentary regime that might have pertained, and to consider whether this regime would

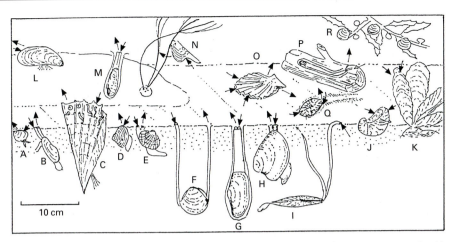

Fig. 5.2 Bivalve life attitudes: (A) *Nucula* (nut shell); (B) *Yoldia*; (C) *Pinna* (fan mussel) with encrusting serpulids; (D) *Astarte*; (E) *Cerastoderma* (*Cardium*) (cockle); (F) *Lucina*; (G) *Mya* (sand gaper); (H) *Mercenaria* (guahog); (I) *Tellina*; (J) *Gryphaea* (devil's toenail), secondary soft-bottom dweller; (K) *Crassostrea* (American oyster); (L) *Mytilus* (mussel); (M) *Pholas* (piddock); (N) *Pteria* (wing oyster); (O) *Pecten* (scallop); (P) *Teredo* (shipworm); (Q) *Glycymeris* (dog cockle); (R) *Posidonia, Bositra* (Fig. 5.4). Arrows indicate incurrents and excurrents. See Fig. 5.21 for rudist bivalves.

have allowed colonization to take place and to be sustained. In the case of a thick turbidite or ash for instance, the speed at which the sediment was deposited means that organisms could only have colonized the top of the bed, though such an 'event' bed may have smothered an earlier community. The best way to determine the sedimentary regime is to examine the stratification (if necessary by cutting and partly polishing a section at right angles to the bedding) and to look for evidence of 'event' beds or of bioturbation. The inferences that can be made from such information are described in Section 6.4.

Normal life attitude Bivalve shells (Figs 5.2 and 5.4), especially infaunal species, are commonly found in life position, as are corals and other substrate-encrusting groups. Look out for pelmatozoan holdfasts or 'rooting' structures. Nestling echinoids between coral branches may be identified by radially arranged spines. Brachiopods (Fig. 5.3) are mainly epifaunal, hence they are less frequently found in life attitude. The mode of life of many extinct groups of brachiopods and bivalves is uncertain. This is particularly true of some of the productid brachiopods, rudist bivalves and thin-shelled bivalves (paper pectens) (Fig. 5.5). Any observations, particularly on autochthonous material are important. The mode of life of the 'paper pectens', a diverse group of thin-shelled, flat-valved bivalves has been much disputed; some interpretations have little observational justification!

Size frequency distribution This would seem to offer clear evidence about authochthony, but there are many factors such as winnowing, additions, spatfall frequency and growth rate, which make it difficult to apply. An occurrence of articulated shells of all sizes is likely to be autochthonous (Section 5.3.11).

Completeness and sorting Leaves and fronds of complex and delicate morphology cannot have been transported far. The extent of skeletal dissociation and valve disarticulation can also be useful indicators of autochthony. Collect for measurement, or measure frequencies of left and right valves, cephala, pygidia, etc. Always consider that large and massive skeletons require the application of considerable force to be shifted. Interpretation of the data can be difficult because, for example, varieties of brachiopod resist disarticulation differently: those with strong tooth socket structure tending to remain articulated. There may also be considerable differences in the breakage potential between valves, as between some pedicle and brachial valves. The number of right valves of oysters is always reduced compared with the number of left valves. Incidentally, free left valves must have been attached to something. It may be possible to identify what this was by looking at the preserved sculpture on the attachment surface. How close are the results to expected frequencies (Appendix B)? With more severe disturbance consider what sorting and winnowing has taken place, and under what type

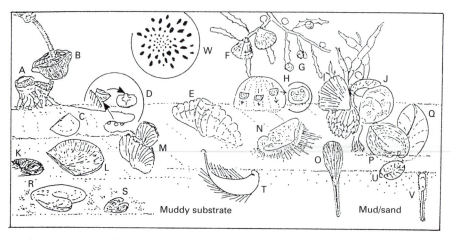

Fig. 5.3 Brachiopod life attitudes (with approximate scales): (A) *Prorichthofenia* (coral-like, umbonal cementation, plus spines) ×0.3; (B) *Linoproductus* (clasping) ×0.3; (C) *Irboskites* (cemented) ×0.5; (D) *Morrelina* (cemented, photonegative thecideidine, brachial valve shown) ×2.0; (E) *Spirifer* ×0.5; (F) *Rhynchonella* (thin-shelled, epiplanktonic?) ×1.0; (G) epiplanktonic inarticulate ×1.0; (H) *Crania* (ventral valve shown, cemented to echinoid test) ×1.0; (I) *Rhynchonella* (pedically attached to alga) ×0.5; (J) terebratulid (pedically attached) ×0.5; (K) spine-stabilized chonetid ×0.3; (L) strophomenid (recliner) ×0.3; (M) orthids (pedically attached) ×0.3; (N) *Waagenoconcha* ('snowshoe' spines) ×0.3; (O) *Terebrirostra* (semi-infaunal) ×0.3; (P,Q) *Conchidium, Pentamerus* (unattached, with umbonal burial) ×0.25; (R,S) *Pygites* ×0.5, *Dicoelosia* ×1.0 (current flow maximization); (T) quasi-infaunal productid ×0.5; (U) free-living terebratulid with thickened umbos) ×0.3; (V) infaunal lingulid ×0.2; (W) *Podichnus* (etch trace of divided, root-like pedicle) ×12.5.

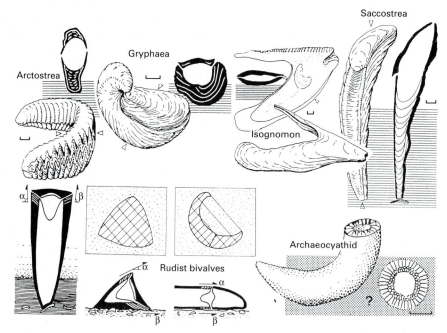

Fig. 5.4 Soft substrate strategies in bivalves and an archaeocyathid. Recliners (low sedimentation rate?): *Arctostrea, Gryphaea, Isognomon*, archaeocyathid (with self-righting 'Savazzi effect'). Mudstickers (high sedimentation rate?): *Saccostrea*, hippuritid (lower left). Cross-sections to show shell thickening between open arrows. Scale bars = 10 mm. Adapted from Seilacher (1984) and Ross and Skelton (1993).

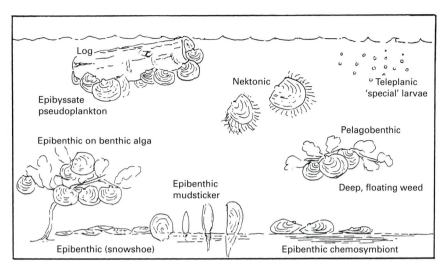

Fig. 5.5 Modes of life that have been suggested for 'paper pectens'.

of hydraulic regime and substrate conditions. Sorting is largely influenced by size, composition and shape; for instance, different skeletal elements of a trilobite behave differently under flow.

Shell breakage and damage The bulk of shell breakage can be attributed to three causes:

- Breakage by hydraulic action prior to deposition. This is of little or no ecological significance, since breakage relates to the form and microstructure of the shell.
- Breakage due to compaction (Section 3.6).
- Breakage due to weathering effects.

Breakage due to compaction and breakage due to weathering effects should be recognizable, since it will be possible to reassemble fragments as they are extracted from the sediment. In younger sediments there may be other causes of shell breakage: predation (shell chipping) by crabs (Fig. 5.8) and predation by birds (probably the main cause of shell breakage on modern tidal flats). Many demersal fish, shell-eating sharks and large crustaceans can fragment shells, but there is no recognizable pattern of breakage.

Attitude and preferred orientation Plotting valve attitude (Fig. 6.5) – convex-up, concave-up, oblique or perpendicular – may give an indication of the hydraulic processes involved (Section 6.4.1). Consider the likely original attitude of the compon-

ents; a low degree of preferred attitude (associated with similar frequencies of valves) would suggest little hydraulic disturbance. For gastropods, absence of preferred orientation in the horizontal plane is the best indication of parautochthony, but overall random orientation and attitude is probably due to bioturbation. Check that the fabric supports this inference.

Organisms respond to a number of orienting influences during life, particularly to light, gravity, currents and food resources (Figs 5.6 and 5.7). Two common examples in the fossil record are burrows along muddy (and more nutritious) ripple troughs, and aligned suspension feeders.

Three-dimensional clustering This can be due to a number of processes. Local pods of shells may be related to burrow infills, gutters or other such structures, load structures, or a portion of a shell bed remaining after erosion. Where the cluster comprises but a few species with many individuals with articulated valves, and some valves deformed because of crowding, then one must have a natural cluster (Fig. 5.8) that was overwhelmed by sediment. Laterally extensive clusters no more than a few centimetres thick may represent local, firm or hardground colonization and may have a very diverse biota. A 'cluster' more or less confined by a concretion (dogger) probably represents early cementation, which allowed the shells to be retained, but diagenetic loss of laterally adjacent shells.

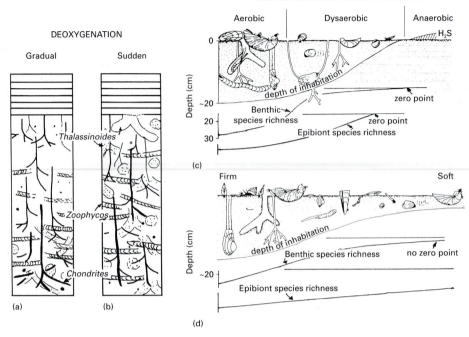

Fig. 5.6 (a, b) Deoxygenation: (a) changes in tier structure with gradual deoxygenation (from observations on North Sea core); (b) a rapid deoxygenation event. (c, d) Distinguishing between soft substrate and low oxygenation: (c) faunal changes under a declining oxygen gradient; (d) faunal changes in a fine-grained clastic substrate on transition from firm to soft conditions. Epibiont species richness refers to hard, shelly substrates. Epifaunal taxa, left to right: *Chlamys*, *Gryphaea* with abundant epibionts including serpulids, *Parainoceramus*, small *Gryphaea*, procerithid gastropod, *Parainoceramus*, serpulid, trochid gastropod, *Gryphaea*, 'paper pecten', *Gryphaea*. Infaunal taxa, left to right: mecochirid crustacean, *Goniomya*, *Nicanella*, *Pinna Nucinella*, *Dentalium*, *Palaeonucula*. There is no zero point in a purely muddy situation. Parts (b), (c) and (d) adapted from Wignall (1994).

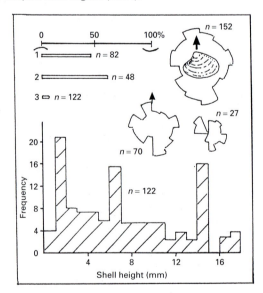

Fig. 5.7 Shell taphonomy in the Kimmeridge Clay. (Upper left) Proportion of convex-up to concave-up valves from three shell pavements. (Upper right) Rose diagrams of orientations of disarticulated bivalves on bedding surfaces; two were from fallen blocks and the third was collected in place. (Lower graph) Size–frequency histogram of *Protocardia* valve height on a single bedding surface; the three peaks probably indicate three successive recruitments of an opportunist. Adapted from Wignall (1989).

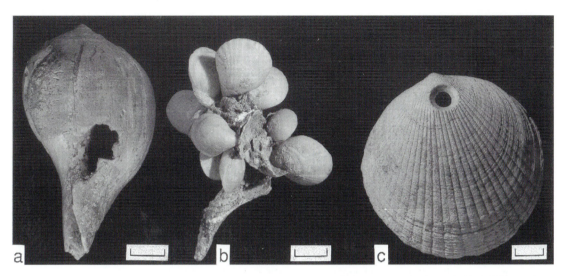

Fig. 5.8 (a) Gastropod (*Sycostoma bulbus*) chipped by a crab (Barton Formation, Middle Eocene, Highcliffe, southern England); (b) terebratulids attached to a gastropod columella (Recent, Australia); *Glycymeris* with drill-hole (*Oichnus*), probably made by a naticid gastropod (Pliocene, East Anglia). Scale bar = 1 cm.

Anomalous presence/absence of selective elements Are there any elements that might be expected, but are conspicuously missing because of selective hydraulic removal, or because of selective skeletal dissolution, e.g. echinoids on an oyster patch where there are indications of the tooth marks (grazing activities) of the echinoids? Or is there the anomalous presence of byssate and encrusting forms on soft substrata?

Faunal composition and winnowing How does the faunal composition of the bed compare with that in adjacent lithologies? Could the fossils have been winnowed from the sediment below?

Encrusters Consider the extent and distribution of encrusters. Does their orientation with respect to the substrate (shell) indicate encrustation during the life of the host (epizoan), as in Fig. 5.20 (see p. 89), or encrustation of a dead shell (epilith)? If it indicates encrustation of a dead shell, has there been subsequent reworking and turning? Encrustation of infaunal bivalves at only the posterior part of the shell may indicate that this portion of the shell was exposed during life, or that penecontemporaneous erosion had partially exposed the shell. Shells with encrusters on both inside and outside must have been on the substrate surface for some time. But are all the shells

similarly encrusted? There may be evidence of several phases of reworking.

Matrix or skeleton support Decide whether the sediment is matrix-supported or skeleton-supported. Matrix support is difficult to explain by hydraulic processes (unless one is clearly dealing with a turbiditic bed, or there has been extensive bioturbation).

Shelly beds

Applications of the criteria to some typical shelly beds are shown in Figs 1.3 and 5.10. Beds analysed in Fig. 5.17 show no shells in life position but varying degrees of dissociation, preferred orientation and left/right valve ratios. Diversity can also be an indication when taken into consideration with other evidence.

Marine mudrocks

Here are some useful criteria for parautochthony in marine mudrocks, i.e. where autochthony is uncertain:

- Matrix-supported shell fabric
- High degree of bioturbation
- Similar frequencies of each valve (in bivalves) even if hydraulically distinct
- Local winnowed horizons
- Many shells convex down, and nesting common
- Little or no preferred orientation

- Lack of abrasion
- Lack of encrusters or borers on infaunal shells

5.2.2 Buildup-producing organisms and their roles

Buildup-producing organisms are diverse, have evolved and been repeatedly replaced over geological time (Fig. 2.6). Many are difficult to identify, especially in the field. Their roles in forming buildups and in producing loose sediment are summarized in Table 5.2, and the relationship of growth form to wave energy and sedimentation rate is depicted in Fig. 5.9. The shapes of stromatoporoids, so characteristic of Silurian and Devonian buildups can be parametrized (Fig. 5.10). The geological literature often refers to distinct development stages in ancient coral and stromatoporoid buildups, but work on Pleistocene and modern reefs (Fig. 5.11) indicates more complex successions (Crame, 1981; see also Section 5.4). In the Jurassic example described in Box 1.1, material has been introduced into the coralliferous unit, and adjacent grainstones carry elements derived from the buildup. Detailed work is necessary to evaluate this 'export–import' model, which Bosence (1995) and colleagues are pioneering on modern south Florida mud mounds and associated sediments.

Table 5.2 Roles of buildup-producing organisms.

Binding, encrusting and stabilizing[a]
Major: laminar rhodophytes, tabulate corals, blue-green algae, oysters, poriferans
Minor: stoloniferous bryozoans, tabulate corals, stoloniferous foraminifera, crinoid holdfasts, sclerosponges, vermetid gastropods

Frame builders (growth fabric)
Primary and macro-sized: lamellar, massive, tabulate corals; strongly branched rugose and scleractinian corals; some bryozoans; rudistid bivalves; some sponges; stromatoporoids; rhodophytes, tubiphytes
Secondary (as primary but smaller scale):[b] diverse groups, especially foraminifera, bryozoa, sponges

Bafflers[c]
Corals, large serpulids, seagrasses, mangroves, fenestellids, archaeocyathids, stalked crinoids, sponges

Producers of loose sediment
Sand-sized, free or weakly linked elements: articulated green and red algae, crinoids, larger foraminifera, sponges, foraminifera attached to seagrasses
Silt to clay size elements: pelagic foraminifera, coccoliths, green algae

[a] Closely attached ± cemented, roots, filamentous, mucilagenous.
[b] Without ability to construct solid frame, but infilling cavities, etc.
[c] Cylindrical ± branching, vase-form.

GROWTH FORM		ENVIRONMENT	
		Wave energy	Sedimentation
	Delicate branching	Low	High
	Thin, delicate, plate-like	Low	Low
	Globose, bulbose, columnar	Moderate	High
	Robust, dendroid branching	Moderate/high	Moderate
	Hemispherical, domal, massive	Moderate/high	Low
	Encrusting	Intense	Low
	Tabular, laminar	Moderate	Low

Fig. 5.9 Growth form of reef-building organisms related to wave energy and sedimentation. Adapted from James (1983).

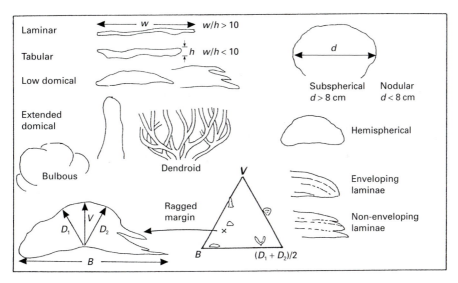

Fig. 5.10 Stromatoporoid morphology and parametrization of shapes (except for dendroid and irregular forms). Adapted from Kershaw and Riding (1978).

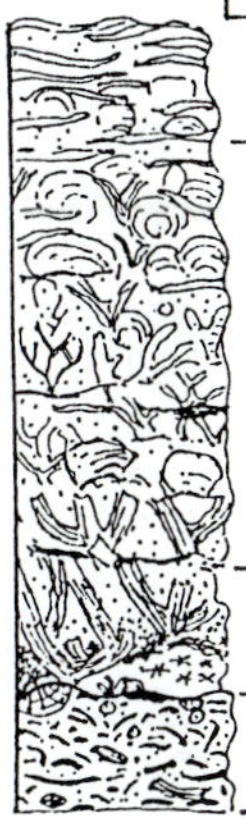

STAGE	LIMESTONE TYPE	DIVERSITY	COLONY SHAPE
Domination	Bindstone–framestone	Low to moderate	Laminar, encrusting
Diversification	Framestone (bindstone) + Wackestone (matrix)	High	Variable: domal massive lamellar branching encrusting
Colonization	Baffle-floatstone wackestone matrix	Low	Branching, lamellar, encrusting
Stabilization	Variable, generally coarse-grained	–	–

Fig. 5.11 Idealized stages in the development of a reef core, and associated types of limestone: species diversity and colony shape. Adapted from James (1983).

5.3 Ecological factors

5.3.1 Ecological gradients

There are three all-important ecological gradients in the world's oceans. Firstly, there is the depth gradient. Light intensity (hence photosynthetic activity), temperature and oxygenation all decrease with depth. Secondly,

there is the shore to open ocean gradient. Nutrient levels tend to be higher in inshore waters either because of upwelling or because of land-derived material. This means that inshore waters have far higher levels of productivity than their offshore equivalents. Thirdly, there is the latitudinal gradient. In low latitudes the seas receive a greater amount of both light and heat than they do towards the poles. Waters at low latitudes therefore tend to be more productive.

These are very broad generalizations and modern seas show that local factors can disrupt expected gradients. For example, nutrient-rich waters upwelling in polar seas can cause very high levels of productivity, or sediment suspended in shallow nearshore waters can greatly reduce expected light levels. It is reasonable to assume that the first of these gradients was as important in the past as it is today, but latitudinal gradients certainly seemed to have varied in intensity. In the Mesozoic, for example, the temperature difference between the poles and the tropics was apparently much less than it is now. The shore to open ocean gradient in the Mesozoic was also probably generally more gentle because higher sea levels gave rise to extensive shallow shelf seas. When making palaeoecological deductions based on these major environmental gradients, it is therefore useful to remember both the importance of local factors and of past sea levels and climates (Box 5.1).

Box 5.1 Eocene sands and muds: shelf to shore

The cliffs at Barton-on-Sea, southern England, are one of the world's classic fossiliferous localities (Fig. 5.12); here the Barton and Headon Formations (middle–late Eocene) are exposed. The rocks dip at a low angle to the east and are close to the axis of the Hampshire Tertiary basin (Figs 1.3 and 1.4). Coastal protection works partly obscure the section. The succession comprises sands and muds representing depositional environments ranging from open marine to estuarine, lagoonal, fluvial and lacustrine. The diverse fauna and flora have been collected for many years and, in part, described in a range of papers and monographs. No modern comprehensive monograph is available. Guy Plint has given an account of the sedimentology and palaeoecology of the upper part of the succession. Two details are provided from a day field class to the section.

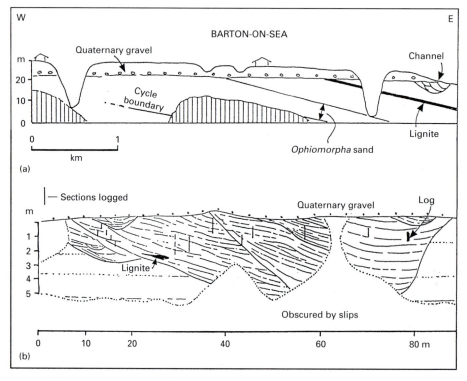

Fig. 5.12 (a) Profile of cliffs at Barton-on-Sea, southern England (for location see Fig. 1.4), with site of cycle boundary, lignite and channel referred to in text. Vertical shading indicates where rock is obscured by sea defences. (b) Sketch of channel with multistage fill, adapted from Plint's undergraduate project report. In the published version (Plint, 1984) areas obscured by slips have been interpreted with full lines. The graphic log illustrated (see Fig. 5.15) is from the most easterly part of the section.

Box 5.1 (cont'd)

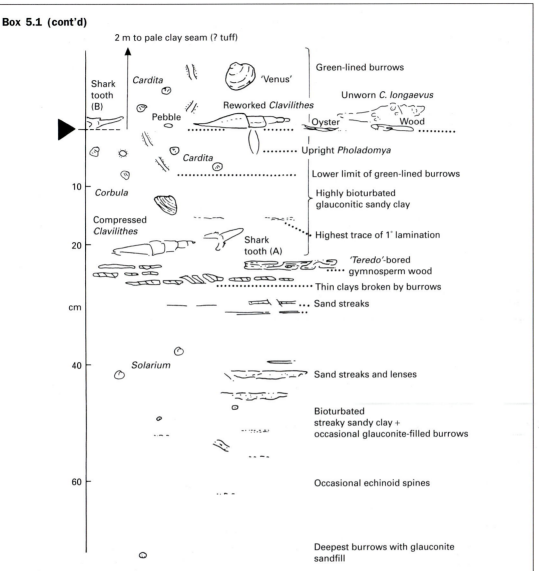

Fig. 5.13 Abbreviated sketch of cycle (sequence) boundary in the Barton Formation (Eocene) at the site indicated in Fig. 1.4. Compiled from field sketches by members of the class, it includes the more common elements of biota and notes on lithologies.

Facies change over a sequence boundary

The object was to investigate the nature of the facies change over a sequence boundary (Fig. 5.13) between two cycles of sedimentation. Guy correlated this with the 42.5 Ma global sea-level event. Due to the low dip and long face, the group of students was able to spread out about 2 m apart. The approximate position of the boundary was readily located because of diggings by local collectors hunting for shark teeth. Scraping with a trowel showed glauconitic sandy clay above, and non-glauconitic fine sand and sandy clay below. But bioturbation was so intense that it took some time to locate the actual junction. Confirmation of the level was indicated by occasional pebbles of flint, small logs (generally '*Teredo*'-bored) and abraded and fragmented shells. The glauconitic sand was seen to be piped downwards with decreasing intensity. Each of the group made a half-scale sketch and the fossils collected were positioned accurately. Some of the fossils were identified at the

Box 5.1 (cont'd)

section using a British Museum handbook, but the diversity was found to be far greater than illustrated. Samples of the sediment were collected from above the boundary in clean plastic bags for laboratory examination of the microbiota and sediment, but sampling below the sequence boundary was difficult because of the bioturbation.

About $1\frac{1}{2}$ hours were spent on the section, and discussion followed about the time represented by the omission event and what evidence might be found to indicate events during the non-represented interval. An aspect that seems to have been overlooked by previous workers is the correlation of shells with overlying and underlying strata. The autochthonous burrowers from the erosion surface and small shells and teeth washed into burrows belong to the overlying sediment; whereas reworked shells from above the boundary belong to the underlying or unrepresented strata. The rare occurrence of brackish water species above the boundary indicates either contemporary allochthonous reworking from a proximal estuarine environment, or that substantial shallowing had taken place during the hiatus. Evidence for erosion having occurred is the presence of worn shells above the boundary with a mode of preservation similar to that otherwise found only below the boundary.

It was noted that many of the shells could be found today in tropical or subtropical locations. This reasoning was supported by the high diversity of the biota and glauconite, which Andrew Hughes (Hughes and Whitehead, 1987) has shown to be primary.

Investigating lignite

About a kilometre to the east, and almost at the end of the traverse, the lignite marking the top of the next sequence was readily spotted. The objectives here were to determine whether the lignite was rooted (Fig. 5.14), and the depositional environment of the underlying sands. It was also intended to investigate other plant-yielding facies to appreciate Guy Plint's work in using the biota in facies interpretation. Incidentally, his detailed investigation arose from a first-year field class when he discovered small calcified roots (anatomically preserved, Fig. 5.14) of a water plant a short distance above the lignite (Crane and Plint, 1979).

The lignite was found to be muddy, with roots slender and prominent. But *Ophiomorpha* (Fig. 8.4) in the underlying sand showed that it had been deposited on shoreface sands and that a very significant drop in sea level had taken place (Fig. 8.28). Most interest was concentrated on the large channel in the cliff above, especially on the final stage of fill,

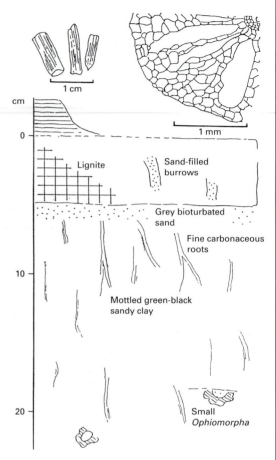

Fig. 5.14 Field sketch of lignite at the base of the Headon Formation along with sketches of calcified (anatomical permineralized preservation) roots of a small aquatic water plant, one with root tip. A sketch from thin section (right) shows typical and well-developed air lacunae. Adapted from Crane and Plint (1979).

which suggested abandonment. Guy based his account on his undergraduate project, from which Fig. 5.12 is taken. Making the sketch of the multi-event channel fill took him some hours of work with a tape and laboriously trowelling out the boundaries. Fig. 5.12 shows the position of the graphic logs he made to gain an appreciation of the facies. The partial log (Fig. 5.15) is new and attempts to show the distribution of the plant-yielding layers. Lignitic logs of gymnospermous wood were common but the vertical face made it difficult to collect good leaf material. With gentle washing, a bag of shelly silt collected from adjacent beds yielded an abundance of shells and gyrogenites of *Chara* (green alga, Fig.

Box 5.1 (cont'd)

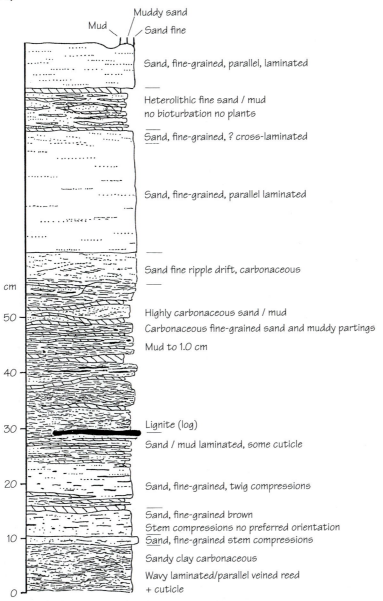

Mud

Muddy sand

Sand fine

Sand, fine-grained, parallel, laminated

Heterolithic fine sand / mud
no bioturbation no plants

Sand, fine-grained, ? cross-laminated

Sand, fine-grained, parallel laminated

Sand fine ripple drift, carbonaceous

cm

50 —

Highly carbonaceous sand / mud
Carbonaceous fine-grained sand and muddy partings

Mud to 1.0 cm

40 —

30 —

Lignite (log)

Sand / mud laminated, some cuticle

20 —

Sand, fine-grained, twig compressions

Sand, fine-grained brown
Stem compressions no preferred orientation

10 —

Sand, fine-grained stem compressions

Sandy clay carbonaceous

Wavy laminated/parallel veined reed
+ cuticle

0 —

Fig. 5.15 Log showing part of the final stage of channel fill with notes on plant fossils.

3.13); this almost certainly indicates freshwater conditions.

Although not yielding particularly good material palaeobotanically, when compared with say Carboniferous sites, this section is important for its record of the major lowstand in sea level (39.5 Ma), and of climatic deterioration during the mid Cenozoic.

Sources: Plint (1984), Crane and Plint (1979)

Table 5.3 Ecological factors.[a]

Physical	Chemical	Biological
Bathymetry	Salinity	Food resources, productivity
Temperature	Oxygenation	Association between organisms
Light conditions	Hydrogen ion concentrations	Abundance
Turbulence		Diversity
Substrate conditions and		Dispersion
geomorphology		Ecological succession
Rainfall		Palaeogeography
Latitude		Birth and mortality rates
Distance from shore (oceanity)		

[a] Many of these factors may experience temporal changes (daily, monthly, annually, etc.).

An appreciation of the role of ecological factors in palaeoecology is best gained before going into the field. Some data are given in Appendix C. The role of many ecological factors may be estimated in ancient sediments by five methods:

1 By comparison with the ecological requirements known for modern taxa, by close comparison for Pleistocene and Tertiary taxa (at species or genus level), or by inferences made on older taxa (generally at genus or family level and with reduced degree of certainty).
2 By making inferences about various morphological features related to specific ecological conditions, e.g. spinosity.
3 By making inferences from sedimentary parameters, e.g. ooids, evaporites, facies relationships.
4 From the samples collected, it may be possible to make determinations by chemical (including isotopic) methods for palaeotemperature and palaeosalinity.
5 By using faunal and floral gradients with respect to the equator, e.g. high diversity and warm temperature at low latitudes.

In general, the precision with which various factors can be determined decreases with geological age. Thus, in the Palaeozoic most determinations are rather speculative, and in the Precambrian largely unrelated to the present; they are anactualistic or non-uniformitarian. It is important to try to use more than one method to determine any parameter. Ecological factors that are of concern to palaeoecology are shown in Table 5.3.

When an appreciation of the nature of the palaeo-environment in question has been gained, it is useful to assess what changes in ecological factors might be expected, bed-to-bed and laterally across the field area

(Table 5.4). This will focus attention on those factors that are of most significance for the particular situation. Any organism relates to a tolerance range and optimum value for each factor. There are also general ecological principles and 'laws', which may occasionally be useful (Table 5.5).

In the field, one's view is generally more limited so far as lateral change is concerned, and it is the vertical, stratigraphic changes that are often of most concern. As mentioned in Section 1.3, Schäfer (1972) recognized the relationship between oxygenation, benthos and stratification. This scheme can be extended to include nutrients and degree of bioturbation (Fig. 5.16).

For terrestrial plants, temperature and precipitation are the most important ecological factors, followed by illumination (length of day, cloudiness) and soil conditions. Indications of these factors are preserved in the plant morphology and soil. Although identification of stomatal and other microscopical characters is a laboratory study, leaf outline and growth rings can be seen in the field. Palaeosols are potentially useful indicators of climate. Most soils, and certainly pre-Upper Palaeozoic soils, do not display evidence of roots (root mould, root casts or petrified roots). Look at the context of the suspect lithology for the likelihood of soil formation, and for evidence of karstification, destratification, colour mottling, calcrete, silcrete, pseudoanticlines, 'non-marine' trace fossils and slickensiding.

5.3.2 Depth and illumination

Apart from shoreline facies, determination of palaeo-bathymetry can only be an estimation. In the field use the following criteria:

Table 5.4 Ecological factors and rate of geographical and stratigraphic change.

	Rate of expected change	
	Geographical[a]	Stratigraphical[b]
Climate	Slow changes only Over extensive areas Minimized during times of globally equable climate	Slow Unlikely over less than a stage Rapid in interglacials
Illumination	High May be local because of turbidity	Rapid if associated with turbidity
Substrate condition	Often very localized	Rapid but reflected in lithology
Bathymetry	If eustatic, likely to affect extensive area If tectonic, rapid and may be quite local If magmatic, rapid and local	Significance decreases with increasing depth Can be rapid in enclosed seas
Turbulence	Generally very localized	Rapid Examples are tidal fluctuations and storm events
Salinity	Local to regional fluctuations Mostly associated with nearshore or lacustrine environments	Rapid Seasonal effects may be over small thickness
Oxygen	Conditions of low oxygenation can be very local or very extensive	Change quite rapid

[a] Geographical changes occur over a wide area.
[b] Stratigraphical changes occur from bed to bed in a section.

Table 5.5 Some ecological principles and laws.

Leibig's law The distribution of a species is determined by that ecological factor which is present in minimum amount. If two (or three) factors approach minimum, survival is especially critical.
Allen's rule Body extremities tend to be smaller in cold climates.
Bergman's principle Species attain larger size in colder regions.
Gause's principle Two species with the same ecological requirements cannot (normally) exist together.

1 Shoreline features, e.g. tidal stratification, herringbone lamination, tidally produced accretion successions, tidal cycles, interference, terraced and truncated ripples, desiccation features: bird's-eye structures, rainprints, footprints.
2 Ooids (generally <5 m, 10–15 m), grapestones (similar but in sheltered areas).
3 Oscillation ripples are generally indicative of depths above normal wave base, but always consider the whole context.
4 Storm event beds (Section 6.3), associated with sedimentary structures such as hummocky cross-stratification, suggest depths to 200 m (storm base), but always consider the whole context.
5 Syndepositional, early diagenetic minerals: glauconite forms today at 125–150 m, chamosite forms today at 10–50 m.
6 Hermatypic corals (<100 m), larger foraminifera (<120 m).
7 Trace fossil bathymetry (Fig. 8.18).

Five other approaches to palaeobathymetry can also be used in the field:

1 Lateral change in facies and in benthonic assemblages.
2 Lateral passage into buildups, which display an ecological succession (providing time lines are available).

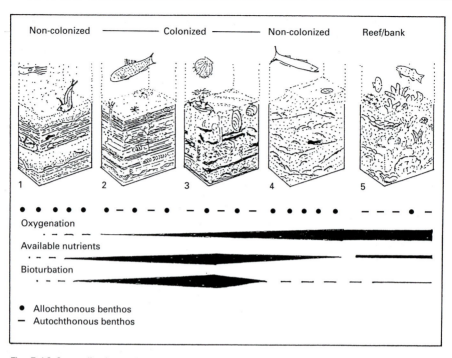

Fig. 5.16 Generalized organism–sediment relationships in marine environments (applicable to aquatic non-marine environments with modification). Autochthonous benthos is absent in (1) and (4), sparse in (2). In (3) the benthos is mainly parautochthonous, and in (5) it is dominantly autochthonous. Oxygenation increases left to right. Nutrients available to organisms are at a maximum in environment (3). The sedimentary record is most complete in environment (1). Bioturbation is restricted in (1) and (4) and most typical of (3). Stratification is essentially absent in environment (5), except for local smothering events (storm and volcanic horizons). In environment (1) there is no benthos. The complete sedimentary record (mud and occasional graded silts) shows an entombed cephalopod, and a concretion has enclosed a fish. Another fish decays at the sediment–water interface. In environment (2) there is sufficient oxygen for a sparse benthos. The sedimentation record is fairly complete. Environment (3) is well oxygenated. The water and sediment (typically muddy sand) carries nutrients for a diverse epifauna and infauna of several trophic types. Turbulence is often high enough to erode and redistribute shells. Environment (4) is too unstable to support much life, and the repeated discontinuities record high turbulence. The almost wholly organic environment (5) is 'reefal' and highly oxygenated. On burial, diagenetic changes tend to obscure the growth forms. Typical situations: (1) euxinic deep basins (Gulf of California); (2) quiet lagoons, deeper oceans; (3) continental shelf, Gulf of Mexico; (4) river mouth bars, oolite shoals, offshore bar sands; (5) modern reefs. The diagram also reflects the not uncommon succession in the fossil record, right to left, of the sedimentary environments found across a marine shelf, nearshore to basin. Adapted from Schäfer (1972), who adopts the following descriptions: (1) letal-pantostrate facies (non-life-complete sedimentation record); (2) vital-pantostrate facies (with life); (3) vital-lipostrate facies (incomplete sedimentation record); (4) letal-lipostrate facies (non-life and incomplete sedimentation record); (5) astrate facies (absence of stratification); most buildups are much more complex (Fig. 5.11).

3 Relief on unconformities and reefs, but remember to correct for compaction.

4 The application of aragonite and calcite dissolution depths (presently 1–2 km for aragonite and 3.5–5.5 km for calcite) depends on the absence of these minerals, e.g. aptychus limestones indicating a depth greater than the aragonite dissolution depth. These depths have fluctuated over geological time. In the Cretaceous they were shallower (calcite ~1.2–2.5 km), subsequently deepening. The Jurassic calcite dissolution depth is estimated at ~3.5 km.

5 Light is not a major factor below 100 m. Blue-green and red algae may extend to this depth (and more), but genera are depth-restricted. Green

algae (utilizing red light) are more restricted, with dasycladaceans found today at <20 m.

5.3.3 Palaeotemperature and palaeoclimate

Palaeotemperature and palaeoclimate can be assessed in the field by several means:

1 Application of the temperature ranges tolerated by extant taxa (e.g. larger foraminifera, thick-walled molluscs and massive scleractinian corals generally indicate warm conditions).
2 Morphological features giving an indication of temperature, e.g. a high proportion of dicotyledonous leaves with serrate margins is typical of humid temperate vegetation, drip tips (tropical, high humidity), leathery leaves and thick cuticles (high-temperature evergreens).
3 Studying the palaeogeographic maps of the area in question to assess palaeolatitude, and thence an indication of climate.
4 Analysis of banding in timber and corals, which is generally associated with seasonality (Fig. 1.2).
5 Do not overlook features of sedimentary origin: desiccation cracks, salt pseudomorphs, frosted and rounded grains, etc.

5.3.4 Type of substrate

The type of substrate principally affects trophism (Appendix C), attachment and penetration of benthic organisms. Criteria for distinguishing between soft (loose), firm and hard substrates are given in (Section 8.5.2). As a generalization, sandy substrates are dominated by filter feeders, and muddy substrates by deposit feeders. In neritic environments the trophic diversity (how many types of feeding) is highest in muddy sand. When looking at fossils that were part of the original benthos, try to determine what the substrate provided for them: (1) a location for attachment (Fig. 5.17), e.g. an epilithic suspension feeder (oyster); (2) protection, e.g. rock-boring clams; (3) a favourable environment for penetration, e.g. rapid-burrowing razor shells; and (4) a source of food, e.g. deposit-feeding, burrowing echinoids.

Sequential changes in biota may be related to taphonomic feedback, i.e changes in community structure due to changes in the nature of the substrate associated with dead skeletal material (Fig. 5.17). There are two types:

- *Facilitative*, e.g. on a soft substrate the accumulation of dead shells provides a substrate for attachment by encrusters (oysters, serpulids) and endobionts (boring bivalves, sponges, worms). The skeletal debris may become the foundation for a buildup, and shell gravels provide refuges and covers.
- *Inhibititive*, where accumulation of dead shells inhibits bioturbation, infaunal burrowers and deposit feeders by elimination or stunting.

Assuming constant shell input:

- Decrease in sedimentation rate leads to greater shell packing density, greater abrasion and fragmentation, bioerosion and encrustation.
- Increase in sedimentation rate leads to reduced shell packing density, less abrasion, fragmentation, bioerosion and encrustation.

5.3.5 Salinity

If one is involved with shoreline or lacustrine facies then salinity is of concern. Several chemical methods are available for the determination of ancient salinities, but in the field only faunal or floral evidence can be used (for trace fossils see Section 8.5.8). These are, in any case, generally more decisive (Appendix C). With rare exceptions, some groups are, exclusively stenohaline (articulate brachiopods, echinoids, calcified bryozoans). Certain nekton (cephalopods) are marine, but marine fish and aquatic reptiles can invade brackish/freshwater systems. Certain taxa are today restricted to freshwater environments (*Viviparus, Planorbis, Unio*, many ostracodes) or will tolerate minimal salinity. Other genera contain species that are normally freshwater and species that are brackish. Size reduction is typical as some bivalves are traced into brackish water, e.g. *Mytilus* in the Baltic. In a lagoonal sequence it may be possible to divide the biota into fresh, brackish and quasimarine associations (Fig. 6.8). Consider the following ideas:

1 Salinity can change rapidly (therefore be on the lookout for rapid stratigraphical change).
2 Freshwater forms can be washed into lagoons and the sea.
3 It is less likely that marine forms have been washed into freshwater.
4 Lagoons and gulfs open to the sea will show a range of salinity across them, but the salinities will not exceed normal marine salinity.

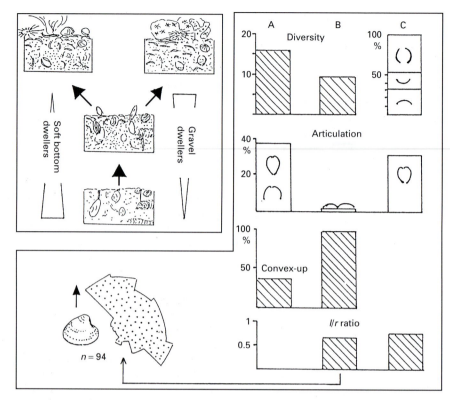

Fig. 5.17 (Upper left) Taphonomic feedback: the accumulation of dead skeletal material may lead to a change in substrate; the accumulation of shells of soft-bottom dwellers leads to an increase in the epifauna, and infaunal gravel dwellers. (Right) Analyses of three shell beds: (A) high diversity, high percentage of articulated shells and relatively low percentage of convex-up orientation indicates parautochthonous storm accumulation; (B) analysis of a shell pavement shows low diversity, very low degree of articulation, most valves convex-up and with preferred orientation, indicating substantial transport; (C) substantial proportion of articulated valves, similar number of right and left valves but poor degree of shell orientation with respect to bedding, suggesting turbulent wave-dominated processes, rather than unidirectional currents, were responsible; the diversity is low. Upper left diagram adapted from Kidwell and Jablonski (1983); right-hand diagrams adapted from Fürsich and Heinberg (1983).

Fig. 5.18 (Left) Crinoid holdfasts and oysters attached to a hardground (Middle Jurassic, Bradford-on-Avon, England); (right) hardground encrusted by *Aulopora*, and bored by *Trypanites* (Devonian, Iowa). Scale bars = 10 mm.

5   Closed or isolated lagoons and lakes can show a range of salinity from zero to halite hypersalinity.

6   The occurrence of freshwater does not necessitate a river supply, since rainfall on surrounding swamps can provide a reservoir.

7   A saline wedge may form in an estuary.

8   Several freshwater taxa are long-ranging and can be traced back to the Mesozoic.

9   Extinct taxa can be used by their association with extant taxa.

10 Primary gypsum and halite pseudomorphs may indicate hypersalinity at overlying levels.

Ascertain the degree of autochthony present. Shell transport in a lagoon will be due mainly to tidal storm and river processes (Box 6.3). Taxa common in estuarine environments are mainly those tolerant to a wide range of salinity (euryhaline). Many are opportunists.

5.3.6   Oxygenation

If there is evidence of hydraulic activity (cross-lamination, cross-stratification, etc.) or evidence of local erosion, it is likely that the substrate was more than adequately oxygenated. The best indicators of anoxia are high organic content (dark colour, bituminous smell) associated with fine lamination (and high lateral persistence) and absence of benthonic organisms, including trace fossils (Section 8.5.7). But there are three things to bear in mind:

1   The oxic–anoxic interface may have been only just above the substrate. Be on the lookout for colonizations of local elevations, e.g. an upper surface of a large ammonite or the upper part of a reptile skeleton.

2   Look out for trace fossils that represent burrowing down from a higher stratigraphic level.

3   The trace fossil *Chondrites* (probably made by a chemosymbiont) by association seems to have been a useful indicator of minimal oxygenation (Fig. 8.4), even though it is usually regarded as eurytopic.

Associated with low oxygenation (dysaerobic) conditions are short-lived oxygenation events (in normally anoxic environments) indicated by a stratum or level of often small bivalves and species tolerant of low oxygen (Fig. 5.7). Wignall (1994) has proposed a general model for marine black shales (Box 5.2). In the field very careful examination of fresh rock is called for. With decreasing oxygenation

this model (Fig. 5.6) predicts changes in size and diversity:

- A decline in diversity of benthic species.
- A decline in organism size (particularly soft-bodied taxa) and hence burrow diameter and depth of penetration.

Candidate oxygenation events are recognized in the field by floods of bivalves, whose mode of life is not without controversy. Certain bivalves occurring in black shales have been interpreted as chemosymbionts. Today such bivalves are all infaunal and include *Solemya*, *Lucina* and *Thyasira* (*Lucina* is illustrated in Fig. 5.2). *Posidonia* (Palaeozoic) and *Bositra* (Mesozoic) have been interpreted in several ways, some theoretically, in the absence of known living examples (Fig. 5.5): (1) pseudoplankton; (2) nektonic (theoretical and very unlikely); (3) teleplanic larval stage (theoretical, planktonic larval stage being retained into adult), but this is unlikely if a morphological change – corresponding to the change from planktonic to benthic (or attached) mode of life – is evident; (4) epibenthic, attached to algae; (5) pelagobenthic attached to floating weed; (6) epibenthic; (7) mudstickers (theoretical), but the morphology is unlike any living mudsticker; (8) epibenthic chemosymbionts. Furthermore, little is known of the density of many woods, especially for the Palaeozoic, and today, *Sargassum* has a high-diversity biota.

Thus, in the field, the shelly biota provide useful indicators of opportunistic colonization during what are often short-lived oxygenation events. They also give useful indications of a very soft substrate (Fig. 5.6). Changes in the size and tier structure of the trace fossils provide the most readily identified indication of deteriorating or ameliorating oxygenation.

5.3.7   Interrelationship

Look for evidence that one organism was influencing another. This is a broad topic; for clustering see Section 5.2.1, for trophic analysis see Section 5.4, and for sexual dimorphism see Section 5.1. In the field keep a lookout for evidence of predation, bored shells, bite marks (Fig. 5.8), that can aid in determining trophic structure. The other common occurrence is of close association. Most encrusting fossils represent no more than the utilization of a hard substrate. For evidence of interdependence it is necessary to show that (1) both components were alive together

Box 5.2 Black shales

Black shales are the world's main petroleum source rocks. They are also associated with good preservation of fossils and represent the more complete sections of the stratigraphic record. Hence they have received much attention from palaeontologists, stratigraphers and sedimentologists. Black coloration indicates high amounts of carbon, but not necessarily anaerobic conditions, and certainly not necessarily deposition at great depth. Their stratigraphic occurrence is somewhat irregular, particularly the thicker units of black shale. In general, thick units of black muds accumulated in epicontinental settings, under low rates of sedimentation. Towards the end of the Mesozoic there is an increase in plankton, especially those types secreting calcareous tests (coccoliths and planktonic foraminifera). The steady rain of their skeletons has perhaps been the main factor in the restriction of black shales. Today they do not seem to be forming to any

great extent and the ocean basins are largely well oxygenated.

Several problems still restrict our understanding of black shales. They include the interpretation of elements of the biota, the degree of oxygenation present, and manner of sedimentation. Paul Wignall has recognized six different facies (oxygen-restricted biofacies) within Jurassic black shales (Fig. 5.19), which range from the abiotic to the fully bioturbated. Fully bioturbated facies have a moderate shelly diversity. The six can be recognized at other stratigraphic levels.

Anoxia is indicated in facies 1 and higher oxygenation levels in facies 6. Reference should be made to Chapter 3; check that an absence of shells is not due to shell dissolution. What has been particularly important in this work is being able to show palaeogeographical variation and temporal trends in the six facies, and how this relates to the geochemical database for the Jurassic black shales.

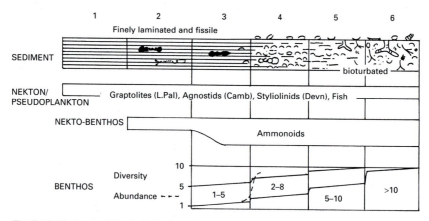

Fig. 5.19 Six types of black shale characterized by sediment fabric and biota. In facies (3) microburrows may be present. Facies (4) is characterized by a marked increase in abundance. Facies (5) has up to ten common species. Facies (6) may have at least ten common species but lacks stenoxic crinoids and corals. Adapted from Wignall (1994).

Source: Wignall (1994)

(Fig. 5.20) and (2) the association was beneficial to one or both. In the Palaeozoic the shelly biota was dominantly epifaunal so that any time surfaces (bedding surfaces) will, ideally, carry the (fallen) elements of a probably tiered association. In Mesozoic and younger sediments more elements were infaunal, with tiering at various levels above and below the substrate.

Fig. 5.21 gives an indication of tiering above and below the substrate through the Phanerozoic for suspension feeders. Complications arise as sediment accumulated with overprinting of successive infaunal levels, or from taphonomic feedback (Section 5.3.4). Analysis can be attempted in the same way as suggested for trace fossils (Section 8.5.6, Fig. 8.24).

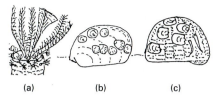

Fig. 5.20 Relationships between shells. (a) True symbiosis between serpulid and hydrozoans (Lower Cretaceous, Gault Clay, Kent). (b) Bivalve (*Lima*) used as substrate for cemented bivalves (Lower Jurassic); the *Lima* could have been alive or dead at the time of colonization. (c) Oysters and serpulids attached to a dead echinoid test (i.e. spines lost). In (b) and (c) the *Lima* and echinoid each acted as a small underwater 'island' for spat settlement. Part (a) adapted from Scrutton (1975); part (b) adapted from Seilacher (1960).

5.3.8 Abundance

Abundance (frequency) is often given as common, rare, very rare, etc. Only the author may know what is meant! If such qualifications are used, try to specify what is meant in terms of numbers per unit area or volume. For instance, trace fossil X might be

described as *rare* if it was found only once over a quarry with 50 beds accounting for 300 linear metres of accessible soles, or *common* if present on one in three soles. Shell abundance can be semiquantified using Schäfer diagrams (Fig. 6.10). For quantified abundance data, collect bulk samples or make counts using quadrats (Section 2.5). With dissociated and fragmented material (the norm for fossils) estimates have to be made, e.g. Table 5.6.

5.3.9 Diversity

Accurate assessment of palaeodiversity depends on what proportion of the original biota is preserved (Fig. 3.2), extractable, determinable and, for the benthos, autochthonous. It is also related to sample size (Section 2.5). There are two components to diversity: richness (the number of species present), and evenness (the relative abundance of each species; see Appendix B.6).

Factors associated with low diversity

High stress This involves any of the physical and chemical limiting factors, e.g. salinity decrease

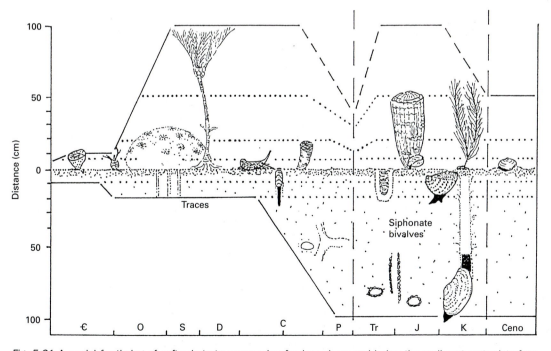

Fig. 5.21 A model for tiering of soft-substrate suspension feeders above and below the sediment–water interface through the Phanerozoic in subtidal shallow marine environments. Dotted lines refer to tier divisions. Adapted from Ausich and Bottjer (1982) and Bottjer and Ausich (1986).

Table 5.6 How to make estimates on dissociated and fragmented material.

Brachiopods and bivalves	*n* of articulated shells + number of either pedicle or brachial valves (or right and left valves) + fraction of broken shells
Gastropods	*n* of apical portions
Cephalopods	*n* of specimens with a complete whorl
Trilobites	*n* of pygidia (or cranidia) *divided by* 10 (number of moults)
Crinoids and ophiuroids	*n* is an estimate of number of stem ossicles *divided by* number (estimate) of stem ossicles per individual
Colonial organisms	number of colonies (for branching colonies, estimate number of fragments that constituted a whole colony)

from open ocean to enclosed sea (e.g. North Sea to Baltic, Indian Ocean to Persian Gulf), turbulence increase towards exposed shore, or biological factors (e.g. predation pressure).

Ecological immaturity   Opportunistic populations may colonize new substrates. Opportunistic (pioneer) species are much overrepresented in the fossil record, and can often be found above event beds, or derived from short-lived environments. It is almost exclusively opportunistic ichnospecies that are found in deltaic facies, even though the swamps were occupied by specialists. Most opportunists are characterized by small size and short life span, thus have few growth lines on shells. In the field look for thin, relatively widespread (but isochronous) horizons, dominated by, especially, one species of body or trace fossil and occupying a low or shallow tier.

Factors leading to increase or maintenance of diversity

Environmental stability   In a stable environment, opportunists are gradually replaced by species with greater capabilities for competitive survival (specialists, equilibrium species of a climax community) and occupying a range of tier levels.

Ecological maturity   Today low latitudes are more stable than high latitudes, and were less affected by the Pleistocene glaciation. Thus the latitudinal (polar) decrease in diversity is especially marked at

present. The deep sea displays high diversity (but low abundance), mainly due to its stability.

Niche availability   Diversity is enhanced where ecological niche availability is high, as on a tropical reef. Similarly, in a tropical area with many islands there is greater opportunity for allopatric speciation to occur.

Factors tending to reduce diversity

Factors tending to reduce diversity are the opposite of those in the previous section: fluctuations in light intensity, salinity (especially in estuarine waters), nutrient supply, etc., and reduction in the number of niches. Diversity is highest where ecological disturbance occurs with just sufficient frequency to prevent competitive exclusion. Shallow-marine allochthonous assemblages may reflect the time averaging of a number of populations or communities (intrahabitat), or a hydraulic concentration from many environments over a vast area (interhabitat). Skeletal material may be transported hundreds of kilometres by storm action.

Things to do in the field

- Sample beds for laboratory investigation.
- Establish trace fossil diversity. For inclusion with body fossil diversity it is necessary to ascertain that the traces belong to the same horizon. Note any correspondence between trace and culprit, e.g. claw in a burrow. Trace and culprit together is rare, e.g. limulid and *Kouphichnium* (see Box 8.2).
- Carefully analyse hardgrounds, buildups or any non-collectable strata.
- Attempt to determine evenness of diversity between assemblages (Appendix B).

5.3.10   Dispersion

When dealing with an autochthonous assemblage there are three possibilities for taxon distribution: random, regular or clumped (Fig. 5.22). For methods of analysis see Appendix B. Clumping can be advantageous for larval settlement or reproduction, but deleterious

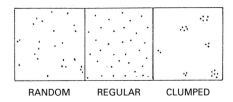

Fig. 5.22 Types of dispersion.

for growth. Regular distributions minimize competition, e.g. in shading or suspension feeding.

5.3.11 Population structure and dynamics

See what information can be obtained in the field about the following items:

1 The age and size distributions of individuals in a population (of a species) – the population structure.
2 Rate of population growth or decline – the population dynamics.
3 Spatial variation in population structure and dynamics, in similar or different facies (see also Section 5.3.10).
4 Long-term changes in population structure.

These questions are the same as an ecologist applies. For any satisfactory synecological study, an attempt must be made to answer these questions. But the problems facing the geologist are many. Consider a few of them:

1 It is seldom that the age reached by a fossil shell can be determined. Nor is growth rate constant over life span. A relationship between age and a growth parameter, such as shell length, will have to be determined and applied. For modern brachiopods, log length is approximately equivalent to age.
2 Consider whether the size distribution of a species in an autochthonous shell bed approximates to the original population structure. It may do so with a census population that was 'caught' below an ash bed or a storm event bed. Otherwise the fossils represent a graveyard. Only if the recruitment and mortality rates were similar will size–frequency histograms approximate to population structure. There is an exception: moults (instars) of benthic trilobites and ostracodes should approximate to a census population.
3 Assess whether an allochthonous shell bed may be used for analysis of population structure. Do not assume that it cannot (Section 6.2).

Generally, one is collecting for later analysis. Collect from (time) intervals that are as narrow as possible. For fossils cemented to bedding surfaces or material that is not feasibly removed, make measurements on individual species of parameters that are age-related (see textbooks), then arrange them in convenient size classes. Ignore compressed and otherwise damaged specimens as far as possible. Analysis is best done having mapped an area or quadrat, so that specimens can be marked as measurement proceeds.

Plotting will also generally be carried out later. But it is useful to have in mind the factors that affect a size–frequency histogram:

- *Biological*: recruitment, including spawning pattern; mortality rate, including seasonal mortality; growth rate; predation rate.
- *Environmental*: substrate type, temperature, etc.
- *Geological*: taphonomic processes, such as sorting, dissolution, fragmentation; collecting failure.

Opportunist (generalist and pioneer) and equilibrium (specialist) species are, respectively, known as r-strategists and K-strategists. Opportunists, with a short life span and small size, tend to be abundant and present in a diversity of facies, whereas equilibrium species are more facies-restricted and with lower growth rate and relatively long life.

An approximation of a survivorship curve (Fig. 5.23) can be constructed from a size frequency distribution. Attempt to determine the relationship between size and age (above). In plotting, the ordinate should be logarithmic (per cent surviving). At size (age) zero, 100% of specimens are 'surviving'. At the 'end' of the first size (age) class, subtract the numbers of that class from the total number of specimens; these are the 'survivors'. To determine their percentage, the calculation for any point on the survivorship curve is

$$\text{Percentage of population surviving at end of class } x = \frac{N - S}{N} \times 100$$

where N = total number of population
S = sum of specimens in classes between 1 and x

5.3.12 Palaeobiogeography

For field observations it is useful to have an appreciation of the principles of animal and plant palaeogeography. What limits a taxon's geographical range? There are just two categories: (1) the chemical, physical and biological limiting factors, and (2) the actual geographical and geomorphological barriers (isthmus, seaway, mountain chain, desert, etc.) that restrict attainment of the potential range. Change in the geographical range of a taxon can only come about by breakdown of geographical barriers (by geomorphological, eustatic or tectonic causes) or changes in a taxon's tolerance to the limiting factors (assuming initial dispersal to be quite rapid). Palaeobiogeography is a complex topic. There are four aspects to consider in the field:

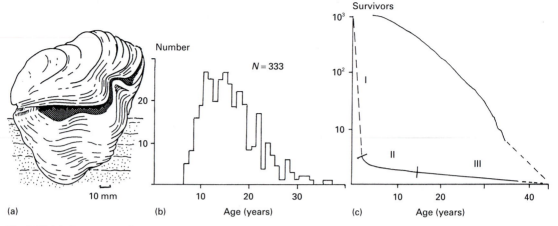

Fig. 5.23 (a) *Crassostrea* cf. *rarilamella* with hinge plates showing; (b) age–frequency histogram; (c) survivorship curves for *C.* cf. *rarilamella*. The upper curve is determined for the actual specimens (no specimens <5 years old were collected). The lower curve is the hypothetical logarithmic survivorship curve divided into three parts: (I) missing first-year class (representing high juvenile mortality), (II) specimens between 6 and 15 years, (III) specimens older than 15 years. Part II of the curve shows a low, but increasing, mortality rate.

Box 5.3 Oyster growth rate and population dynamics

During field mapping of Lower Tertiary sediments in the southern Pyrenees (Spain), Allard Martinius, of Utrecht, found a bed of the boulder-shaped oyster (suspension feeder) *Crassostrea* cf. *rarilamella* (Fig. 5.23). The oysters were in life position in muddy estuarine sediments and presented an opportunity for determining the population dynamics of the colony, though Allard recognized that taphonomic processes have decimated the high diversity of the biota found on modern oyster banks.

After cleaning some 350 specimens collected randonly from ten 4 m² surfaces, the annual growth increments were counted and measured over the hinge plates, and computer analysed. Living *Crassostrea* is a much studied taxon and this enabled Allard to make particularly useful comparisons. The distribution of age class may be considered as normal since 95% of individuals have an age at death in the range of 5–28 years. But high juvenile mortality has considerably skewed the histogram, giving it a large positive skew. The survivorship curve indicates high juvenile mortality and constant adult mortality, with low decrease of the adult population. And *C.* cf. *rarilamellosa*, with its high maximum age and adaptation to muddy sediment, appears to have been a K-strategist.

Source: Martinius (1991)

1 Absence of taxa that might ordinarily be expected to be present, or vice versa.
2 Differences in composition across tectonically linked regions (Section 7.7.3).
3 The apparent migration of taxa across an area because of shift of climatic belts (but this is only applicable to areas that are hundreds or thousands of square kilometres in extent).
4 Presence (rare) of a taxon that usually occurs in another province, e.g. a warm-water (Tethyan) ammonoid in a cooler (Boreal) province. The occurrence may be due to chance floating-in following death.

5.4 Palaeocommunities and palaeoenergetics

At quite an early stage in an investigation of the palaeoecology of a fossiliferous site, it is useful to attempt a pictorial reconstruction of the site as it was at the time of formation. This focuses one's mind wonderfully on asking the pertinent questions. Later a pictorial reconstruction of the palaeocommunity can be made, which can be fully justified from the analyses.

Many substrate reconstructions have been published, but few have been based on a quantitative assessment of the sedimentology and palaeontology. Many are no more than inspired guesses. Of course,

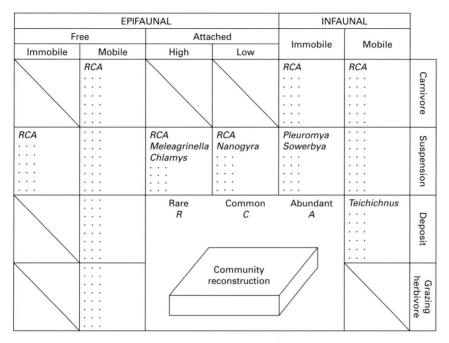

Fig. 5.24 Trophic substratum mobility chart and community reconstruction, with an example from the Jurassic depicted in Fig. 1.5.

any reconstruction will depend to a considerable extent on how *community* is defined. A useful and practical definition for palaeontologists is that a *palaeocommunity* is an association of organisms that previously lived together. This places emphasis on assessment of autochthony. A definition that will involve more sophisticated analysis is that a palaeocommunity is an association of organisms which, when alive, formed an interdependent association. This will involve statistical tests and determinations of former biomass and energy flow. Further discussion is beyond the scope of this section, but it is useful to consider the major problems confronting analysis of fossil material: (1) that energy flow can only be determined on the basis of an individual's life span, not on a daily/annual basis (as in modern ecology); (2) the relationship between shell size (all that remains) and the former biomass is not readily determined or estimated; (3) no adequate allowance can be made for the soft-bodied biota; (4) taphonomic loss was probably greater than it appears to be.

As work progresses on modern substrates, especially on shelf benthos, we are gaining a better appreciation of how modern communities can be defined and how they might become fossilized. Particularly relevant to ancient sequences is the appreciation of

short period fluctuations, and long-term changes in population structure. One might expect these fluctuations and long-term changes to be well represented in the fossil record, in the development of fossil reefs, in changing successions on soft substrates, and so on. But it is not easy to prove naturally occurring 'autogenic' successions, where the succession is biologically controlled, though not necessarily deterministically; and in most cases, physical and chemical processes have intervened, thus leading to community replacement.

A simple method of analysing fossiliferous sediments is to construct a trophic substrate mobility chart (Fig. 5.24) followed by a sketch of the substrate reconstruction. An indication of relative abundance can be included, together with an indication of the tiering above and below the substrate of body and trace fossils. Feeding habits and habitats can also be presented on ternary diagrams (Fig. 1.5). Plotting should be done as species abundance rather than species presence, because some species may be represented by only a few specimens. Use of such diagrams has been criticized because, when applied to modern communities, analysis of the dead (shelly) fauna (analogous to the fossil material) is distorted when compared with that of the live fauna. Similarly suspect is the use of trophic nucleus (number of species that account for 80%; Fig. 1.5), but both may be useful in facies analysis. If there are sufficient data, more sophisticated statistical analysis can be applied to discriminate the palaeocommunities.

5.5 Fossil-ores (Fossil-Lagerstätten)

There are a number of fossiliferous localities known to just about every geologist: the Solnhofen Limestone with *Archaeopteryx*, the Burgess Shale with many soft-bodied animals and trilobites with preserved appendages, or the Rhynie Chert with the exquisite preservation of early vascular plants. There are many others. Such sites (Fig. 5.25) are often known as Fossil-Lagerstätten or sites of extraordinary preservation. Many result, however, from the coincidence of normal taphonomic processes with normal sedimentological situations and biological evolution. A careful sedimentological analysis of the fossiliferous site may reveal the processes that gave rise to the preservation.

Using the term *fossil-ore* reminds us that the mining geologist locates ore bodies on a scientific basis by mapping, prospecting and predicting where to recommend mining with a defined degree of confidence. And, of course, certain sulphide and other types of ore deposits represent the sites of ancient marine hydrothermal vents (smokers), with the extraordinary chemosynthesizing biota of giant gutless worms, giant clams and predator crabs. The hot springs of Rotorua, New Zealand, were taken as a modern analogue for the gold-bearing Lower Devonian cherts of Rhynie, Scotland, by Trewin (1994); and the waningstage volcanism of the Yellowstone area, United States, was taken as a model for similar deposits (Carboniferous) in Queensland, Australia, by Walter *et al.* (1996), both exhibiting a perfection of plant cell preservation, as well as primitive arthropods at Rhynie.

There are five main categories of primary fossil-ore:

1 Many autochthonous buildups (reefs, Section 2.6), and hardgrounds (Section 8.5.2).
2 Occurrences characterized by an abundance of fossil remains (fossil concentration deposits) due to hydraulic processes, sorting or winnowing material, e.g. winnowed bone concentrates, richly fossiliferous cross-stratified gravels, or fissure fills representing hydraulic traps (ecological sieves) – sometimes called *Concentrate Lagerstätte*.

3 Occurrences characterized by articulated skeletons (especially of vertebrates and echinoderms). The most important factor is absence of bioturbation and reworking, hence seek deposits representing rapid covering, rapid burial by submergence in muddy 'liquid' sediment (deforming any lamination), incorporation into a high-density flow (wackestone) – Burgess Shale situation – or amber, or deposits indicating an anaerobic substrate. Most large marine vertebrates are found in mudrocks. Expect differences in the ability of live animals to have escaped from such situations (low for echinoderms and trilobites), and variation in the way dead material may have responded.
4 Occurrences characterized by the presence of soft tissue as well as completeness (Section 3.3.1) – sometimes called *Conservation Lagerstätte*. These are more common than is generally thought and include bioimmuration (Box 3.3). Again, absence of bioturbation and reworking is critical. In older Palaeozoic strata, deep burrowers were only present locally in nearshore sands. The beautifully preserved Cretaceous flying insects and occasional pterosaur of Brazil were pickled in lacustrine brine.
5 Occurrences associated with vents, seeps and hot springs. There are three types (Gaillard *et al.*, 1992):
  • Hydrothermal vents associated with oceanic crust, and massive sulphide deposits. H_2S is the dominant source for chemosynthesis. Known from Cyprus, Oman and New Caledonia.
  • Cold seeps are associated with continental crust and subduction zones. Methane is the dominant nutrient source. Known from France, Canada and northern Italy.
  • Continental hot springs associated with late-stage volcanism (see above).

A crinoid complete with stems and arms attached to the calyx, or a trilobite virtually complete, or a pocket of well-preserved brachiopods, is a minor fossil-ore. Pose the appropriate question: Might there be a brittlestar on the sole of this storm bed? Might there be insect remains associated with all these plant fossils?

Secondary 'fossil-ores' are where fossils are reworked from any of the major types of fossiliferous sediment (Chapter 1) and concentrated by modern processes. An example is the wave concentrate along Kent (southern England) beaches of pyritized twigs, seeds and fruits derived from the cliffs of London Clay (Eocene), a type 5 fossil deposit. Another is the strip-mine dumps at Mazon Creek (Illinois) with fossiliferous nodules from a type 10 fossil deposit.

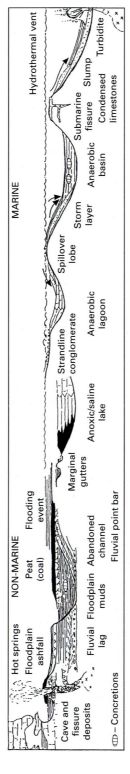

Fig. 5.25 Environmental settings of some primary fossil ores. Adapted from Seilacher *et al.* (1985).

Looking at fossiliferous sites in this way helps to give a better appreciation of the fossil record. But with many fossil-ores there is a price to be paid. Most have a strongly biased biota, e.g. general absence of benthos at Solnhofen, loss of aragonitic fossils in the Faringdon Sponge Gravels. All the more important are the rare occurrences where a high proportion of the original assemblage has been preserved, as occurred in the Silesian of Mazon Creek (Nitecki, 1979), in the Middle Cambrian Burgess Shale (Whittington, 1985) and in the Eocene of Messel.

References

Ausich, W. I. and Bottjer, D. J. (1982) Tiering in suspension-feeding communities on soft substrata throughout the Phanerozoic. *Science* **216**: 173–74

Bosence, D. W. J. (1995) Anatomy of a Recent biodetrital mud-mound, Florida Bay, USA. In Monty, C. L. V., Bosence, D. W. J., Bridges, P. H. and Pratt, B. R. (eds) *Carbonate mud-mounds: their origin and evolution.* Special Publication 23, International Association of Sedimentologists, pp. 475–93 (Blackwell Science publishes some IAS items)

Bottjer, D. J. and Ausich, W. I. (1986) Phanerozoic development of tiering in soft substrate suspension-feeding communities. *Paleobiology* **12**: 400–20

Crame, J. A. (1981) Ecological stratification in the Pleistocene coral reefs of the Kenya coast. *Palaeontology* **24**: 609–46

Crane, P. R. and Plint, A. G. (1979) Calcified angiosperm roots from the Upper Eocene of southern England. *Annals of Botany* **44**: 107–12

Fürsich, F. T. and Heinberg, C. (1983) Sedimentology, biostratinomy and palaeoecology of an Upper Jurassic offshore sand bar complex (Greenland). *Bulletin of the Geological Society of Denmark* **32**: 67–95

Gaillard, C., Rio, M. and Rolin, Y. (1992) Fossil chemosynthetic communities related to vents and seeps in sedimentary basins: the pseudobioherms of southeastern France compared to other world examples. *Palaios* **7**: 451–65 (Part of a thematic set of papers on chemosynthetic communities)

Hughes, A. D. and Whitehead, D. (1987) Glauconitization of detrital silica substrates in the Barton Formation (Upper Eocene) of the Hampshire Basin, southern England. *Sedimentology* **34**: 825–35

James, N. P. (1983) Reef environment. In Scholle, P. A., Bedoid, D. G. and Moore, C. H. (eds) *Carbonate depositional environments.* American Association of Petroleum Geologists, Tulsa OK

Kershaw, S. and Riding, R. (1978) Parameterization of stromatoporoid shape. *Lethaia* **11**: 233–42

Kidwell, S. M. and Jablonski, D. (1983) Taphonomic feedback: ecological consequences of shell accumulation. In Tevesz, M. J. S. and McCall, P. L. (eds) *Biotic interactions in recent and fossil benthic communities.* Plenum, New York, pp. 195–248

Martinius, A. (1991) Growth rates and population dynamics in *Crassostrea* cf. *rarilamella* from the Lower Eocene Roda Formation (southern Pyrenees, Spain). *Geologie en Mijnbouw* **70**: 59–73

Nitecki, M. H. (ed.) (1979) *Mazon Creek fossils.* Academic Press, New York

Paul, C. R. C. and Bockelie, J. F. (1983) Evolution and functional morphology of the cystoid *Sphaeronites* in Britain and Scandinavia. *Palaeontology* **26**: 687–734

Plint, A. G. (1984) A regressive coastal sequence from the Upper Eocene of Hampshire, southern England. *Sedimentology* **31**: 213–25

Powell, E. N. and Stanton, R. J. (1985) Estimating biomass and energy flow of molluscs in palaeo-communities. *Palaeontology* **28**: 1–34

Ross, D. J. and Skelton, P. W. (1993) Rudist formations of the Cretaceous: a palaeoecological, sedimentological and stratigraphic review. *Sedimentology Review* **1**: 73–91

Schäfer, W. (1972) *Ecology and palaeoecology of marine environments.* Oliver & Boyd, Edinburgh

Scrutton, C. T. (1975) Hydroid–serpulid symbiosis in the Mesozoic and Tertiary. *Palaeontology* **18**: 225–74

Seilacher, A. (1960) Epizoans as a key to ammonoid ecology. *Journal of Paleontology* **34**: 189–93

Seilacher, A. (1984) Constructional morphology of bivalves: evolutionary pathways in primary versus secondary soft bottom dwellers. *Palaeontology* **27**: 207–37

Seilacher, A. (1991) Self-organizing mechanisms in morphogenesis and evolution. In Schmidt-Kittler, N. and Vogel, K. (eds) *Constructional morphology and evolution.* Springer, Berlin, pp. 251–71

Seilacher, A., Reif, W.-E. and Westphal, F. (1985) Sedimentological, ecological and temporal patterns of fossil Lagerstätten. *Philosophical Transactions of the Royal Society of London B* **311**: 5–26

Trewin, N. H. (1994) Depositional environment and preservation of biota in the Lower Devonian hot-springs of Rhynie, Aberdeensbire, Scotland. *Transactions of the Royal Society of Edinburgh: Earth Sciences* **84**: 433–42

Walter, M. R. *et al.* (1996) Lithofacies and biofacies of mid-Palaeozoic thermal spring deposits in the Drummond Basin, Queensland, Australia. *Palaios* **11**: 497–518

Whittington, H. B. (1985) *The Burgess Shale.* Yale University Press, New Haven CT

Wignall, P. B. (1989) Sedimentary dynamics of the Kimmeridge Clay tempests and earthquakes. *Journal of the Geological Society, London* **146**: 273–84

Wignall, P. B. (1994) *Black shales.* Oxford University Press, Oxford

Further reading

Ager, D. V. (1963) *Paleoecology.* McGraw-Hill, New York

Benton, M. J. (1997) *Vertebrate palaeontology: biology and evolution.* Chapman & Hall, London

Bignot, G. (1985) *Elements of micropalaeontology.* Graham & Trotman, London

Boardman, R. S., Cheetham, A. H. and Rowell, M. (eds) (1987) *Fossil invertebates.* Blackwell, Oxford

Brasier, M. D. (1980) *Microfossils.* George Allen & Unwin, London

Brenchley, P. J. and Harper, D. A. T. (1998) *Palaeoecology: ecosystems, environments and evolution.* Chapman & Hall, London

Carroll, R. (1987) *Vertebrate palaeontology and evolution.* W. H. Freeman, New York

Clarkson, E. N. K. (1998) *Invertebrate palaeontology and evolution,* 4th edn. Chapman & Hall, London

Dodd, J. R. and Stanton, R. J. (1990) *Paleoecology: concepts and applications,* 2nd edn. J. Wiley, Chichester

Gray, J. (ed.) (1988) *Palaeolimnology: aspects of freshwater palaeoecology and biogeography.* Elsevier, Amsterdam (Book edition, with index, of *Palaeogeography, Palaeoclimatology, Palaeoecology* **62**: 1–623)

Gray, J. S. (1981) *The ecology of marine sediments.* Cambridge University Press, Cambridge

Lehmann, U. (1976) *The ammonites: their life and their world.* Cambridge University Press, Cambridge

Lipps, J. H. (1992) *Fossil prokaryotes and protists.* Blackwell, Oxford

Raup, D. M. and Stanley, S. M. (1978) *Principles of palaeontology,* 2nd edn. W. H. Freeman, New York

Retalleck, G. J. (1990) *Soils of the past: an introduction to paleopedology.* Unwin Hyman, London

Schopf, T. J. M. (1980) *Paleoceanography.* Harvard University Press, Cambridge MA

Skelton, P. (ed.) (1993) *Evolution: a biological and palaeontological approach.* Addison Wesley, London

Smith, A. (1984) *Echinoid palaeobiology.* George Allen & Unwin, London

Stewart, W. N. and Rothwell, G. W. (1993) *Paleobotany and the evolution of plants,* 2nd edn. Cambridge University Press, Cambridge

Tavesz, M. J. S. and MCcall, P. L. (eds) (1983) *Biotic interactions in recent and fossil benthic communities.* Plenum, New York

Taylor, T. N. (1981) *Paleobotany: an introduction to fossil plant biology.* McGraw-Hill, New York

Thomas, B. A. and Spicer, R. A. (1986) *The evolution and palaeobiology of land plants.* Croom Helm, London

Whittington, H. B. and Conway Morris, S. (eds) (1985) Extraordinary fossil biotas: their ecology and evolutionary significance. *Philosophical Transactions of the Royal Society of London,* Vol. 192 (A collection of papers covering fossil ores of types 3 and 4)

Wray, J. L. (1977) *Calcareous algae.* Elsevier, Amsterdam

Wright, V. P. (ed) (1986) *Paleosols: their recognition and interpretation.* Blackwell, Oxford

Shell concentrations and skeletal elements as hydraulic and environmental indicators

Without fossils, sedimentology is lifeless and timeless.

John Pollard

This chapter is chiefly concerned with shells and shell concentrations. Shell concentrations have been largely the preserve of palaeontologists for palaeo-community analysis and in seeking morphological or stratigraphical information (Chapters 5 and 7). The skeletons of plants and animals can form accumulations in life position (as in the case of coral reefs or other buildups) or can be accumulated by non-biological sedimentary processes. We are concerned here with the latter type, but consider that reefs may often show areas of allochthonous accumulation as a result of factors such as storm damage and winnowing. Another special problem concerns accumulations formed by organisms such as some calcareous algae and the larger foraminifera (*Nummulites*, orbitoids, fusulines), which may still be able to survive if disturbed (Box 6.1).

6.1 Information potential and general strategy

Shell concentrations can provide much information that relates to the hydraulics of the sedimentary environment (and which may fluctuate with time), and hence to facies interpretation, and also to correlation and sequence stratigraphy. Before setting about an analysis of a concentration, the overall context of the units should be viewed and logged. Shell bed analysis is likely to take time and it is important to ask questions about what one aims to achieve:

- What information can be obtained about the hydraulic regime that operated before deposition was initiated, and the regime during deposition? Do they record normal or event (short-lived) activity?

- How much erosion took place (if any) at the base of the bed, and what does this means?
- What happened, successively, through a unit? Is more than one event represented (see Fig. 6.2, p. 101)? Is the top graded? Or was it recolonized? If so, was it, by burrowers (Chapter 8), by encrusters or by a subsequent shelly biota (taphonomic feedback)?
- Where through a bed are the skeletal materials concentrated?
- What are their orientations? Is there preferred orientation? If so, what does this mean?
- What is the composition of the materials? Ask all the questions that relate to the palaeoecology and taphonomy of the materials.
- What happens to the unit as it is traced laterally? How might it terminate? Does this give clues about the depositional environment?
- What diagenetic information can be obtained? Might this be significant for the overall diagenetic history of the unit, e.g. the history relating to porosity and permeability: retention of cavities for gas entrapment in large foraminifera, or aragonite dissolution providing space for oil.

Taphonomic feedback (Fig. 5.17) needs to be considered if a shell bed is recolonized and/or bioturbated. It then becomes a composite bed and each part must be analysed separately, even if the sections are 'welded' together, when the bioturbation may present problems. For instance, gutters in Silurian neritic muddy sediment, filled mainly with dissociated crinoid ossicles acted as a substrate for coral colonization and growth (and in some cases for metre-thick buildups). In a Mesozoic situation the top of a shell bed was colonized by bivalves, different from those in the main part of the bed, and then the bed was disrupted by bioturbators.

Terminology

With the current interest in shell concentrations it is not surprising that the terminology is rather fluid. Do not force field observations into one or other classification. Section 6.3 describes 11 main types of shell concentrations, together with several minor types. Some represent *event* conditions, others are composite (multi-event) conditions. Kidwell (1991, with earlier papers) and Brenchley and Harper (1997) recognized only four types. Here are the main differences between the definitions. Their *event* concentrations include recolonization; and their *lag* concentrations are restricted to multiple events, where there was substantial erosion at the base of the unit, and components are second-cycle (derived from usually lithified sediment). Lags, as considered here, represent (shell) accumulations *which normal hydraulic activity did not entrain*, e.g. shells accumulating in a laterally eroding stream (Fig. 3.6).

6.2 Attributes of shell concentrations

Those that reflect the final depositional event

- The overall geometry and extent.
- The nature of the bounding surfaces, e.g. erosive ± gutters.
- The thickness of the unit.
- The primary sedimentary structures present, e.g. hummocky cross-stratification ± graded bedding.
- The nature, close-packing and size-sorting of the skeletal components.
- The orientations of the skeletal elements (primary, secondary, etc.).
- The nature of intraformational clasts and derived fossils.
- The nature of the matrix (wackestone, packstone, grainstone).

Those that reflect the preceding history

- The composition of the biological materials, and the interpretation of their original life environment; the palaeoautecology and palaeosynecology: relative abundances of taxa, diversity (species, genus, monotypic, polytypic), ecological spectrum (benthos, nekton), age spectrum.
- The taphonomic history of the materials (disarticulation, fragmentation, abrasion, bioerosion (boring, biting), encrusting (Fig. 5.18).

Those that affected the succeeding history

- The post-depositional history of the unit as a whole.

Decide which of these attributes is most important for the job in hand. They can be estimated or determined semiquantitatively. The best way to make a semiquantitative determination is to have a checklist to hand and to go through it for each element. Charts can then be drawn up or a software program used. Analysis will be quicker as skill is gained. Consider sample size (Section 2.5) to minimize counts. Use a ruler or calipers for size determination, and for complete shells select which parameter will be recorded (e.g. diameter, length). If using non-quantitative methods, then high/medium/low must be specified, e.g. by visual comparators (Fig. 6.10, see p. 110).

6.3 Types of shell concentrations

Fig. 6.1 shows the distribution of 11 main types of shell concentrations that Fürsich and Oschmann (1993) recognized across an onshore–offshore marine gradient, together with some additional but common concentrations, set against the distribution of the processes considered to be responsible. Also included are shell concentrations found in marginal and deep-marine settings. For each, analyse whether it represents a single event or is composite, and the extent, thickness, fabric (in plan and side profile) and the taphonomy. For reference these shell concentrations are listed in the same order as given by Fürsich and Oschmann (1993).

(1) Fair-weather wave concentrations

- *Multiplicity*: Event or composite.
- *Extent*: Generally not greater than a few 100 m.
- *Thickness*: Each event, less than 1 m.
- *Fabric*: Bioclast supported; poor sorting; bimodal plan orientation.
- *Taphonomy*: Fragmentation high; encrusting, boring absent (unless shells derived from buildup or rocky coast).

(2) Storm wave concentrations

- *Multiplicity*: Event or composite.
- *Extent*: Generally less than 100 m.
- *Thickness*: Most typically cm to dm thickness.
- *Fabric and taphonomy*: Erosive base; packstone–grainstone; bimodal sorting (complete shells and comminuted debris); random orientation (plan); hummocky cross-stratification (shells in pods).

SKELETAL CONCENTRATIONS

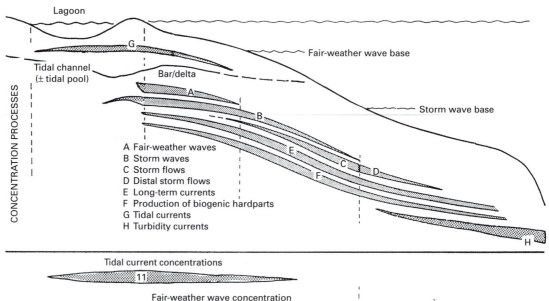

CONCENTRATION PROCESSES

Lagoon

G

Fair-weather wave base

Tidal channel
(± tidal pool)

Bar/delta

A

Storm wave base

B

A Fair-weather waves
B Storm waves
C Storm flows
D Distal storm flows
E Long-term currents
F Production of biogenic hardparts
G Tidal currents
H Turbidity currents

E

C D

F

H

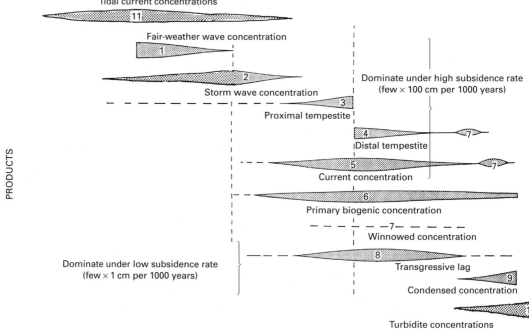

PRODUCTS

Tidal current concentrations

11

Fair-weather wave concentration

1

2

Dominate under high subsidence rate
(few × 100 cm per 1000 years)

Storm wave concentration

3

Proximal tempestite

4

7

Distal tempestite

5

7

Current concentration

6

Primary biogenic concentration

7

Winnowed concentration

8

Dominate under low subsidence rate
(few × 1 cm per 1000 years)

Transgressive lag

9

Condensed concentration

10

Turbidite concentrations

Fig. 6.1 A process-related classification of skeletal accumulations with 11 types of accumulation along an onshore–offshore gradient, grouped according to subsidence rate. Based on Fürsich and Oschmann (1993) and data from the Jurassic of Kachchh, western India, and Kidwell (1993) and her study of the Salton Trough (Neogene), Gulf of California (high rate) and Miocene sediments of the US Atlantic (passive) margin (low rate). Lens thickness reflects relative frequency.

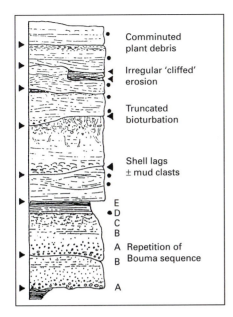

Comminuted plant debris

Irregular 'cliffed' erosion

Truncated bioturbation

Shell lags ± mud clasts

E
D
C
B
A   Repetition of
B   Bouma sequence

A

Fig. 6.2 Criteria for recognizing amalgamation in turbidites (below) and storm beds (above). Arrows indicate first-order discontinuities at amalgamations; spots indicate second-order discontinuities within or between flow regimes.

Storm beds and tempestites are particularly common in the rock record, but they are something of a ragbag! Fürsich and Oschmann (1993) distinguished between storm beds where evidence of lateral transport is insignificant, and tempestites (below) where deposition involved extensive lateral transport. Evidence for transport must come from the make-up of the skeletal elements (their inferred life environment, relative to that of the tempestite), and also non-biological clasts. For instance, tempestite deposits in the southern North Sea contain *Hydrobia*, a small gastropod derived from backbarrier tidal flats.

Storm beds and tempestites are often composite. There is often an indication of storm frequency (relative) if the time represented by a member or formation can be estimated. Composite beds (Fig. 6.2) may, at least in part, represent the cannibalization of beds formed earlier. Nearly always storm beds are cut out when traced laterally by penecontemporaneous erosion. It has not been established whether 'belts' corresponding to 'storm tracks' occur. Successive units of composite beds may show different compositions and taphonomic histories. The distribution of shells in storm beds is generally as pockets, for instance, in the troughs between hummocks. In decalcified beds these have to be carefully sampled or blocks should be taken for acid digestion. Mark and label the samples.

If the field evidence is interpreted as a storm or tempestite unit, but a unit lacking shells, ask the question why? Could provenance or shell dissolution be an explanation?

Distinctions between storm beds, tempestites and turbidites have not yet been fully investigated. Probably the most useful differences that can be seen in the field are in the manner of recolonization, as seen by the ichnology (Section 8.5):

- *Storm beds*: colonization represented by suite of trace fossils identical to that of background sediments.
- *Tempestites*: colonization probably as for storm beds.
- *Muddy tempestites*: no colonization or sparse colonization.
- *Turbidites*: colonization by a different suite of animals, leading to a different suite of trace fossils.

Awareness of the might of *tsunami* is increasing. Their effect and products on land have been described but not the type of bed formed on the marine shelf or in deeper water. Lateral extent should be greater than for tempestites.

(3) Proximal tempestites

- *Multiplicity*: Most commonly composite.
- *Extent*: Greater than for storms; probably up to several kilometres.
- *Thickness*: Commonly a metre thick.
- *Fabric*: As for type 2 but with indications of transport; packstone to wackestone ± grading ± cross-lamination/hummocky cross-stratification, flute casts.
- *Taphonomy*: Disarticulation general.

(4) Distal tempestites

- *Multiplicity*: Generally event.
- *Extent*: May be several kilometres.
- *Thickness*: Generally not greater than centimetres.
- *Fabric*: Thinner than type 3; sorting better than type 3; shells convex-up ± grading.
- *Taphonomy*: Fragmented > complete valves; abrasion high.

(5) Current concentrations

- *Multiplicity*: Generally event.
- *Extent*: Varies from stringer, gutter fill to sheet of kilometre extent.
- *Thickness*: Single shell to cm–dm thickness.
- *Fabric*: Cross-stratification (in thicker units), packstone matrix.
- *Taphonomy*: Abrasion and disarticulation high ± bioerosion.

Table 6.1 Distinctions between wave- and current-formed shell accumulations.

	Waves	Currents
	LINEAR SHELL ACCUMULATIONS	
Distance between shell bands	Uniform, as ripple wavelength	Variable
Frequency of shell bands	Always numerous	Often isolated, rarely numerous
Shell attitude within bands	Long axes parallel to bands	Imbrication in one direction
Deposition of small biogenic particles	On each side of band	Only on lee side
Direction of gutter casts	Transverse to bands	Parallel to bands
	EDGEWISE SHELL ACCUMULATIONS	
Direction of long axes	Different orientations	Uniform (±), at right angles to current
Inclination	Various directions	Constant, generally upcurrent
Accumulations	Stacks in various directions	Imbrication in one direction (downcurrent)

Source: Futterer (1982).

Generally seen on bedding surfaces, *stringers* (Table 6.1) are linear arrangements of shells, including ostracode valves, coprolites and faecal pellets; they are quite common in muddy sediments. Stringers can be useful in giving evidence of higher-energy events in muddy sediment more or less devoid of coarse siliciclastic grains. Two types can be distinguished in the field. Type 1 (more common) is current-formed, akin to current lineation, type 2 is wave-formed, akin to starved (isolated) ripples. On modern beaches with wave ripple, fine and light particles tend to accumulate in the ripple troughs.

There is probably a gradation from stringers to *pavements* (Fig. 5.17) formed by a layer of shells one valve thick, or no more than two valves thick. Pavements (plasters) are also common in mudrocks. The preferred orientation and possible imbrication of the elements will be a clue to transport direction. Check that the size distribution and shell attitude does not oppose interpretation as a hydraulic concentration (Section 5.2.1).

Any *flute casts* may be filled with shells. Fine skeletal material or silt and fine sand may accumulate in the lee of a larger shell to form a *current shadow*.

Bedding expression must also be studied. Generally seen normal to the bedding, *gutter fills* (casts) define storm track and possibly depositional slope; they are usually centimetres wide (rarely decimetres) and centimetres deep. The soles are generally rounded in cross-section. Note whether the gutters are straight or sinuous, and whether groups are present. Are there minor groovings on the sides? Is the top rippled? If so, by what type of ripple? Is there more than one episode of cut and fill? Transport direction will probably be apparent only by slabbing. Could the fill material have been derived from the adjacent facies? Stringers, pavements, current shadows and gutter fills are illustrated in Fig. 6.3.

(6) Primary biogenic concentrations

Primary biogenic concentrations result from the gregarious habit of the organisms and are covered in Chapter 5.

(7) Winnowed (washed) concentrations

- *Multiplicity*: Generally a single episode.
- *Extent*: Probably not greater than 100 m.
- *Thickness*: Generally <10 mm, but see Box 6.1.
- *Fabric*: Stratification poor, sole-graded; look for lateral passage to sharp sole.
- *Taphonomy*: Shell completeness often high (including thin shells); abrasion minor ± encrustation.

Winnowed accumulations describe the gradual removal of finer particles by current or wave action. The essential question to ask is whether the sole is sharp or gradational. Winnowing will leave a gradational sole. But in practice, by following the level laterally, a site can often be found where actual erosion occurred and the shells lie on a sharply truncated surface.

(8) Lags in general

Lags are generally composite. The commonest type of lag deposit is the coarse material at the base of a hydraulically transported bed of sand. The basal conglomerate to a transgression surface may also be included. There will always be an erosion surface below the lag, and this distinguishes it from a winnowed or washed accumulation. Imbrication may be evident,

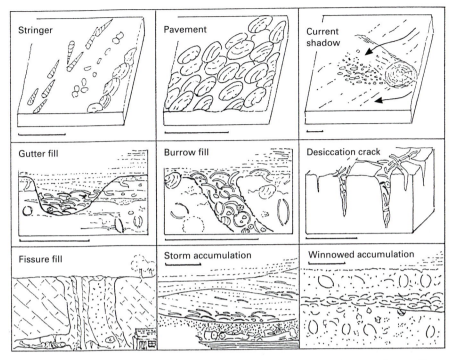

Fig. 6.3 A process-related classification of skeletal accumulations: all except fissure fill are rapidly formed, event beds (scale bar = 0.1 m). Adapted from Kidwell *et al.* (1986).

and cross-lamination may be discernible, depending on the shapes of the materials. Reworked material will generally have been through early stratinomic and diagenetic processes; for instance, the pores of bony or echinoderm elements become infilled with mineral matter or clay-grade sediment, thereby much increasing the specific gravity: fresh bone 2.0, fossil bone 2.6–2.9. Such material is known as *prefossilized*. 'Beach' lags generally show high lateral continuity, whereas lags formed during lateral migration (e.g. meandering streams) may show frequent discontinuities with rapid lateral change in composition.

Transgressive lags

- *Multiplicity*: Composite.
- *Extent*: Can be extensive, many kilometres but then probably diachronous.
- *Thickness*: Can be up to a metre thick.
- *Fabric*: Often complex ± amalgamation and reworking; omission overlap bioturbation; incipient cementation; random orientation (overall); evidence of different hydraulic processes.
- *Taphonomy*: Disarticulation, fragmentation, abrasion variable through unit; encrustation, boring high.

(9) Condensed concentrations

- *Multiplicity*: Composite.
- *Extent*: May be extensive, especially at maximum flooding of basin.
- *Thickness*: Generally cm–dm.
- *Fabric*: As type 7 but longer time involved ± mixed zonal fossils if at sequence boundary, or transgression surface, but less overall transport ± evidence of early diagenesis and reworking of concretions, sole less regular than type 6; pockets of shells.
- *Taphonomy*: Time averaging high ± shell dissolution.

(10) Shell concentrations in turbidites

- *Multiplicity*: Event or composite.
- *Extent*: Can be up to thousands of kilometres.
- *Thickness*: Centimetres to metres.
- *Fabric and taphonomy*: As for tempestites.

(11) Tidal current concentrations

- *Extent*: Local to tens of metres in channels, more extensive in point bars (see type 8).
- *Thickness*: Centimetres to metres.
- *Fabric*: Fragmentation variable.

- *Taphonomy*: Variable, depending on derivation of materials.

Tidal current concentrations may be regarded as a 'special' case of current deposit, since it is rare to find evidence of bimodality (ebb and flood flows). Some minor types of shell concentrations include:

- Shell clusters and patches
- Burrow fills
- Desiccation structures
- Fossils found in tectonic fissures
- Relict accumulations

Shell clusters and patches (Section 5.2.1) are local parautochthonous accumulations due to decay of the tissues connecting skeletal elements as with a stalked crinoid, or the seaweed to which brachiopods or other shells were attached (Fig. 5.6). They indicate mimimal hydraulic activity or scavenging, etc.

Burrow fills arise in two ways: either shelly material is washed into the burrow, tubular 'tempestites' of Tedesco and Wanless (1991); or material is selected out by the burrower and biogenically pressed into the backfill. Some burrows have a purpose-built wall of shell material, e.g. tube worms *Lanice* and *Diopatra* (Fig. 8.30).

Desiccation structures may become shell-filled. Check that desiccation was the process forming the crack, by determining the configuration of the crack and its plan. (Syneresis cracks made underwater are generally much narrower and incomplete, but may also fill with shells, e.g. ostracodes and shell fragments.)

Of larger dimension are solution fissures, which mostly result from subaerial, karstic processes. When submergence has followed, they may not be easy to distinguish from sedimentary (Neptunian) dykes of tectonic origin. Search for features indicative of a subaerial origin. Terrestrial caves and fissures may yield richly and they may be extensive.

Fossils found in tectonic fissures are always of interest, because they often represent events for which no stratiform record is preserved. Relict accumulations of skeletal debris are widespread today on continental shelves (Fig. 2.1). Following the rapid eustatic rise of the last 10 000 years, and the changed conditions of light, food supply and low net sedimentation, the dead shells and skeletons remain on the seafloor, or accumulate over a long period. Ancient accumulations of relict material may be difficult to prove, but there is likely to be evidence of prolonged exposure on the seafloor: encrustation, extensive borings (post-Silurian) and carbonate dissolution.

6.4 Analysis of biofabric

The fabrics listed for the various types of shell accumulations in Section 6.3 are for a rapid visual assessment. When the aim is to gain an insight into the original structure of the palaeocommunity, a more detailed analysis is required (Table 6.2). The extent to which this might be achieved will depend on several factors:

- The degree to which the bed can be broken up without further damage to the components. With lithified sediment it will often only be possible to use decalcified material and then to make an assessment of the bedding-parallel surfaces or aspects in cross-sections.
- Taphonomy affects different types of skeleton in different ways (Chapter 3) and the distribution of the processes is in part environment-related (Fig. 6.4).
- The composition of a shell bed seldom represents a single event. Winnowed beds probably represent the most likely candidates. Most beds are multicyclic and the shells will have been repeatedly reworked. This need not be too serious since the *time-averaging* of storm beds can still provide good indications of the primary shelly community.

6.4.1 Reorientation and preferred orientation

Fossils show preferred orientation (Fig. 6.5) for one of two reasons:

- They are in life position and took up their attitude under the influence of light, gravity or nutrient-supplying currents.
- They were oriented after death by wave or current action, or biogenic agencies.

What is generally required is an indication of hydraulic strength and direction or trend. Preferred orientation is best measured on bedding surfaces. Measure the angle to the dip direction with a protractor, or from a non-digital watch laid parallel to the dip. Measure trends separately (0–180°), e.g. trilobite axes and brachiopod hinge lines, and directions (0–360°), e.g. gastropod apices and brachiopod commissures. For plotting rose diagrams use 10°, 15°, 20° or 30° intervals depending on the number of readings. With dips greater than 25°, correct the vector mean for tectonic tilt using a stereonet (or program). Note that current or wave motion needs to act over a period of time to

Table 6.2 A semiquantitative approach to the characterization of skeletal accumulations.

Feature	Scale
TAPHONOMIC FEATURES	
Articulation	articulated → disarticulated but closely associated → disarticulated and dissociated
Size sorting	v. good → mod. good to poor → v. poor → bimodal
Modal size and range	use φ or mm scale
Shape sorting	
Fragmentation (type of breakage)	whole → fragments
Abrasion (polish)	unabraded → abraded → highly abraded/polished
Rounding	angular → subangular → rounded → well rounded
Orientation	all in life orientation → mixed → all disturbed
in plan view	preferred orientation → high variance
in cross-section	preferred orientation → high variance
SEDIMENTOLOGICAL ATTRIBUTES	
Matrix type	
Hydraulic equivalence of matrix	same as shells → more mobile → less mobile
Relative abundance of elements	see Fig. 6.10
Close packing of elements	
Associated sedimentary structures	
PALAEONTOLOGICAL ATTRIBUTES	
Number of species	
Relative abundance	
Taxonomic composition	
Ecological spectrum	nekton, plankton, benthos, infauna, epifauna, etc.
	soft-, firm-, hard-substrate dweller
Age spectrum	
Original mineralogy	
Preserved mineralogy	

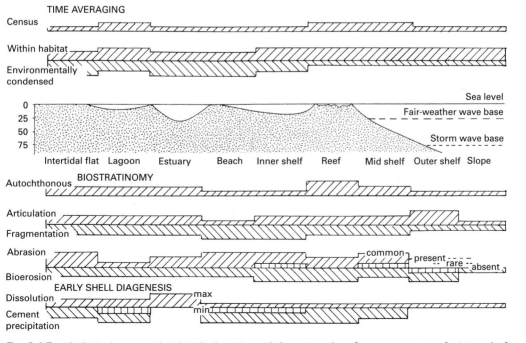

Fig. 6.4 Trends in taphonomy related to bathymetry and time averaging. Census: a near perfect record of live, shelled taxa. Within habitiat: modified record due to post-mortem disturbance and/or time averaging. Autochthonous: in position of life. Adapted from Kidwell and Bosence (1991).

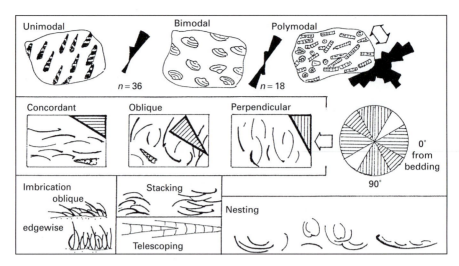

Fig. 6.5 Terminology for shell orientation on bedding surfaces and in cross-section: unimodal, unidirectional current; bimodal, wave oscillation; concordant/convex down, settling from suspension; perpendicular/edgewise, oscillatory flow; oblique imbrication, upcurrent dip; stacking/ nesting, storm reworking. Adapted from Kidwell *et al.* (1986).

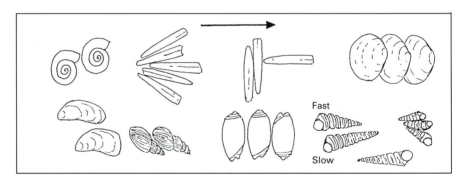

Fig. 6.6 Orientation of some shells in a current. Adapted from Futterer (1978).

induce shells into the position of least resistance. A firm bed is also required; on a soft/loose substrate the elements tend to become buried.

The orientation of shells, etc., in currents is reasonably predictable from their shape. Fig. 6.6 shows various types of shell and their orientation under flume conditions. Flattish shells tend to form imbricate arrangements readily. Tall spiral gastropods rotate about the apertural end, with the apex tending to point downstream, except under slow flow. Surprisingly, even coiled cephalopods may become oriented with the aperture more or less downcurrent, and heteromorphs even more so. Strong wave action (seen on modern beaches) leads to edgewise orientation (Fig. 6.5, Table 6.1, Box 6.1).

Small shells are not uncommonly concave-up in association with small-scale cross-lamination. Elongate particles (wood, crinoid stem lengths) may roll or align parallel to the current. A good indicator is a T-arrangement (belemnites, crinoid stems) with the crossbar being upcurrent. More complex arrangements of crinoid stems and the 'screw' bryozoan *Archimedes* have been observed in the field and reproduced in the flume. Belemnites may show complex patterns, because they tend to roll along a curved path. In muddy sediments, compaction may make it difficult to perform orientation measurements.

Animals with flexible or protruding parts, such as starfish and brittlestars, have been found below a smothering sediment with the arms drawn out – a

Box 6.1 Nummulite accumulations

The large foraminifer *Nummulites* and other taxa are the main rock formers of large petroleum and gas reservoirs in North Africa, the Middle East and the western Pacific regions. Like modern reef-forming corals, they almost certainly had photosymbionts, and probably inhabited shallow oligotrophic seas. Flume experiments show that the ovoid to discoid skeletons were readily disturbed and reworked by wave and current action. While investigating quarries originally worked for limestone blocks to build the pyramids, Tom Aigner proposed a model to explain the alternations of poor and well-sorted nummulites (Fig. 6.7(a)). He suggested that periods of

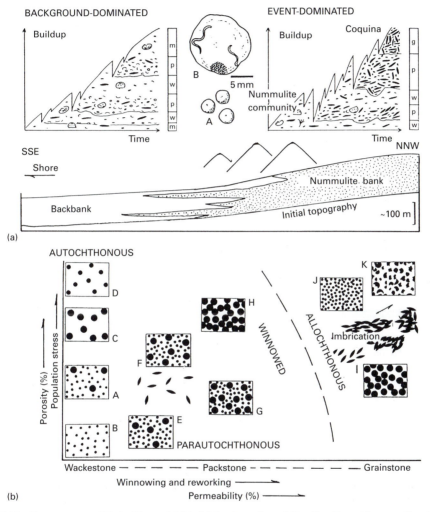

Fig. 6.7 Middle Eocene nummulitic buildups. (a) Model for formation of the Giza Pyramids area, Egypt, to show distribution of nummulite bank and adjacent backbank facies. Left and right: generalized facies logs to show contrasting biofabrics. Bank, with alternating autochthonous and winnowed accumulations, building upwards to high-energy grainstones (g) (coquinas). Backbank mudstones, (m) wackestones (w) and packstones (p), forming in deeper water, with fewer nummulites and a mixed benthos and packstones, the latter indicating periods of storm activity. (b) A dynamic model for the distribution of nummulitic biofabrics in Oman. (A–D) Autochthonous fabrics formed under different conditions: (A) 'normal' conditions, (B) 'most favourable' conditions, (C) unfavourable conditions and (D) unfavourable conditions with stunted B-forms (or where B-forms reproduce early). (A normal assemblage has A : B = 10 : 1). (E) Parautochthonous biofabric with mud winnowed away. (F–H) biofabrics produced in different ways: (F) winnowing and enrichment in B-forms, (G) further enrichment and (H) removal of A-forms. (I, J) transport and sorting, leading to packstones and grainstones: (I) with B-forms (I) and (J) with A-forms. (K) nummulithoclastic biofabric of fragmented nummulites and other skeletal materials, a fabric found especially in forebank facies. Part (a) adapted from Aigner (1984); part (b) adapted from Racey (1994).

Box 6.1 (cont'd)

autochthonous growth were interrupted by winnowing episodes.

As Andrew Racey working in Oman has shown (Fig. 6.7(b)), this model is rather too simple. The protozoans produced two types of individual. The smaller A-form is produced asexually (with a larger initial chamber), whereas the larger B-form (up to several centimetres diameter) is sexually produced and has a small initial chamber. The original ratio of A-forms to B-forms in a 'normal' environment is thought to have been in the ratio of A : B = 10 : 1. But under stress conditions only B-forms were produced or stunted/juvenile B-forms preserved.

Exceptionally favourable conditions led to only A-forms. Thus, before any winnowing might have taken place, there could have been four types of primary accumulation! Logging showed successions from autochthonous 'normal' accumulations to 'stressed' assemblages, indicating upward-increasing stress as the buildup came under greater wave turbulence. Racey recognized autochthony by dominance of mudstone and wackestone lithologies, rare abrasion, dominance of A-forms, encrustation by small oysters or red algae (often on one side only), borings relatively common, and a complete gradation in test size within A- and B-forms.

Sources: Aigner (1984), Racey (1994)

---

rare occurrence. In contrast, the more common tangled mass of crinoids around a log of wood in Jurassic mudrock indicates low energy.

Rapid sedimentation is generally associated with a low degree of preferred orientation, poor sorting, high frequency of convex-down valves and, possibly, high numbers of articulated shells. Note that bioturbation may also lead to the first three of these attributes.

6.4.2   Derived fossils

Derived fossils may not always be readily identified as such. They provide clues about the source sediment, which may be very local, and possibly about local tectonics. Their distribution regionally will give clues about the palaeogeography (Box 6.2).

6.4.3   Matrix

Matrix is hydraulically significant only if it can be shown to be primary, and not due to infiltration (giving non-uniform distribution and umbrella effects) or due to compacted pellets, the identity of which is often preserved in shelter cavities (Fig. 2.9). Collect material for laboratory slabbing. Distinguish between clast-supported (packstone, grainstone) and matrix-supported (wackestone) fabrics.

6.4.4   Diagenesis

Diagenesis may modify an assemblage considerably. In coarse, loose sediment, aragonitic shells may be completely lost, or there may remain only rare indications of their original presence. In contrast, the

Box 6.2 Derived fossils

Jon Radley, Andy Gale and Mike Barker reported derived Jurassic marine fossils from the lagoonal Lower Cretaceous of the Isle of Wight, southern England, and explained the concentrations by analogy with processes operating in a modern temperate lagoon with a Jurassic shoreline! Phosphatized and worn Jurassic ammonites and bivalves are well known from the Lower Cretaceous and readily recognized as such. But when one is searching for clues to salinity, a respectable echinoid radiole (spine) or small oyster can be temptingly persuasive evidence! And when such fossils occur with some frequency, the temptation may be too great! It is just such a situation that makes Jon, Andy and Mike's paper so relevant. In addition to radioles and oysters, they report marine foraminifera, serpulids, several bivalves and encrusters and borings. These were derived from Jurassic strata undergoing erosion along an active fault-scarp cliff, probably not more than 2 km distant from the section. This gives useful tectonic information. The concentrations of derived fossils occur in and at the top of shell beds as storm concentrations swept from the old shoreline, and as winnowed concentrates.

Source: Radley *et al.* (1998)

number of (originally) aragonitic shells may be relatively enhanced in a shell bed where early cementation has taken place, whereas only rare specimens remain in life position in an underlying leached siltstone.

6.5 Shell concentrations in basin analysis

Box 6.3 describes the shelly biota of a stratified Cretaceous lagoon. Certain types of shell beds (Fig. 6.9)

occur in a moderately predictable position in stratigraphic successions, and in parasequences:

1 At a lower sequence boundary, lag concentrations are common. The boundary unit may be quite thick (>1–2 m) and composite, with a mixture of lag and tempestite units.

Box 6.3 A stratified Cretaceous lagoon

Ancient lagoonal sediments are widespread and offer tempting insights into an environment that is relatively easily appreciated today. The ecological parameters that influence the shelly biota are complex. Franz Fürsich and James Kirkland's analysis of the sediments and shells in a 2.3–3.4 m thick succession in the Cenomanian of Arizona (Fig. 6.8) portrays a stratified water body with a local saline wedge below storm wave base, which was only colonized by lucinoid bivalves. The parautochthonous shell beds and pavements contain an abundant brackish water suspension feeder (*Caryocorbula*). But detailed analysis shows there were four other taxa that

became locally common, though not exclusively so. By carefully evaluating the environmental parameters controlling faunal distribution (energy level, grain size, substrate consistency, sedimentation rate, food availability, salinity, oxygenation), Franz and James concluded there was not just one controlling parameter but several, and they varied throughout the unit. Thus their model was for low-salinity waters within the reach of storms, and a capping with rather lower salinity in a lagoon with a prograding shore. The lagoonal sediments were eventually overlain by transgressive and reworked barrier-bar sands.

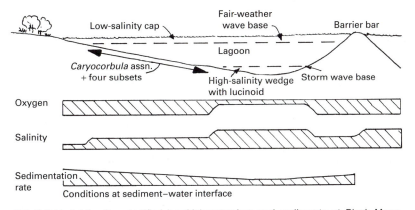

Fig. 6.8 Model for the stratified brackish-water lagoonal sediments at Black Mesa, Arizona. Adapted from Fürsich and Kirkland (1986).

Source: Fürsich and Kirkland (1986)

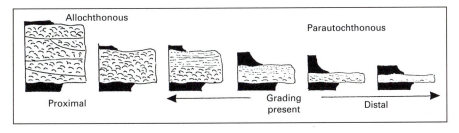

Fig. 6.9 Proximal to distal sequence in a shell bed. Adapted from Fürsich and Oschmann (1986).

2 The maximum flooding unit is characterized by condensed deposits, and winnowed beds are typical of the more muddy units.

3 Tempestites and storm beds tend to be concentrated in the highstand systems tract.

4 Wave concentrations may occur in the upper part of the highstand systems tract associated with progradation.

5 A lag is often present at the base of each parasequence, with an upwards succession from winnowed to distal to proximal tempestites or storm bed.

6.6   Quantitative description

The quantitative description of a skeletal accumulation does not generally in itself have a great deal of hydraulic significance. The parameters of size, shape, packing and sorting, as well as diversity, depend on the nature of the original associations as well as hydrodynamic processes. Also important are sequential changes (vertically and laterally) in these parameters. Fig. 6.10 shows diagrams to give an

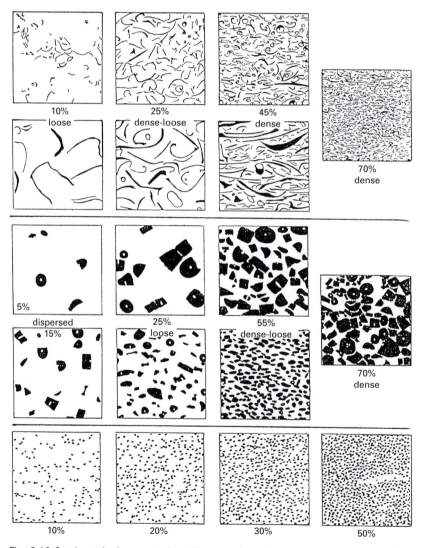

Fig. 6.10 Semiquantitative comparison diagrams for estimating percentage of skeletal material (shells, crinoid elements) and ooids in a sediment. To be used at ×2. Categories of close packing (Fig. 6.11) have been added. Adapted from Schäfer (1969).

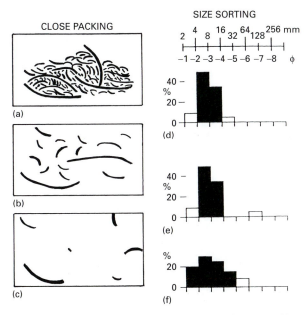

CLOSE PACKING

SIZE SORTING

Fig. 6.11 Schematic illustration of categories of close packing and size sorting for bioclasts >2.0 mm. (a) densely packed, bioclast supported; (b) loosely packed, a few clasts in contact, most within a length of each other; (c) dispersed, clasts well separated; (d) well-sorted, 80% lie within one or two adjacent φ size classes; (e) bimodal, well-sorted but with distinct second mode; (f) poorly sorted, over 80% lie within three or more size classes. Nine combinations can be used, e.g. loosely packed and well-sorted. Adapted from Kidwell and Holland (1991).

assessment of a skeletal accumulation, comprising shells, crinoid elements and ooids (or rounded grains); they can be used separately or in combination. The matrix should be described separately. Categories for close packing and size sorting are illustrated in Fig. 6.11.

6.7 Tool marks

Tool marks (Figs 4.3, 4.4 and 6.12) are the result of objects, such as mud clasts, together with skeletal and plant material (fragmentary or in more complete condition), being rolled or saltated by hydraulic processes (current or wave action) over the substrate surface and locally impinging on the surface. The marks frequently leave a sense of flow direction or wave trend and thus may be useful when other indications are absent. Tool marks are seen on upper surfaces and as casts on soles. They may be grouped as roll or tumble marks, brush marks (elongated with a small mound at the downcurrent end), bounce marks (symmetrical), prod marks (asymmetrical, broadening downcurrent), skip marks (repeated impact) and drag marks (much elongated).

Current action on a rooted weed can lead to arcuate marks (Fig. 4.3). Only occasionally is it possible to identify the exact nature of the tool. The high degree of preferred orientation means that tool marks are seldom confused with trace fossils. Biological objects, such as the tubes of emergent worms, may interfere with and destabilize current flow, leading to local scouring. Pressing a piece of soft plasticine onto casts helps in appreciation of the mark.

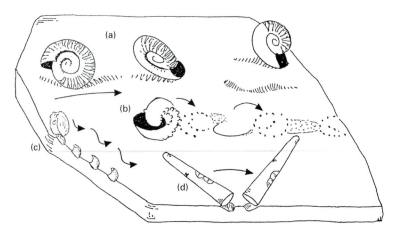

Fig. 6.12 Tool marks: (a) roll marks made by a swaying perisphinctid ammonite on a fine-grained substrate; (b) tumbling marks made by an inflated tuberculate ammonite; (c) skip marks made by a vertebra; (d) orthocone prod mark. Parts (a) and (b) adapted from Seilacher (1963); parts (c) and (d) adapted from Dzulynski and Walton (1965).

References

Aigner, T. (1984) Facies and origin of nummulitic buildups: an example from the Giza Pyramids Plateau (Middle Eocene, Egypt). *Neues Jahrbuch für Geologie und Paläontologie, Abhandlungen* **166**: 347–68

Brenchley, P. J. and Harper, D. A. T. (1997) *Palaeoecology: ecosystems, environments and evolution.* Chapman & Hall, London

Dzulynski, S. and Walton, E. K. (1965) *Sedimentary features of flysch and greywackes.* Elsevier, Amsterdam

Fürsich, F. T. and Kirkland, J. I. (1986) Biostratinomy and paleoecology of a Cretaceous brackish lagoon. *Palaios* **1**: 543–60

Fürsich, F. T. and Oschman, W. (1986) Storm shell beds of *Nanogyra virgula* in the Upper Jurassic of France. *Neues Jahrbuch für Geologie and Paläontologie Abhandlungen* **172**: 141–61

Fürsich, F. T. and Oschmann, W. (1993) Shell beds as tools in basin analysis: the Jurassic of Kachchh, western India. *Journal of the Geological Society, London* **150**: 169–85

Futterer, E. (1978) Studien über die Einregelung, Anlagerung und Einbettung biogener Hartteile im Strömungskanal. *Neues Jahrbuch für Geologie und Paläontologie Abhandlungen* **156**: 87–131

Futterer, E. (1982) Experiments on the distribution of wave and current influenced shell accumulations. In Einsele, G. and Seilacher, A. (eds) *Cyclic and event stratification.* Springer, Berlin, pp. 175–79

Kidwell, S. M. (1991) The stratigraphy of shell concentrations. In Allison, P. A. and Briggs, D. E. G. (eds) *Taphonomy: releasing the data locked in the fossil record.* Plenum, New York, pp. 211–90

Kidwell, S. M. (1993) Taphonomic expressions of sedimentary hiatuses: field observations on bioclastic concentrations and sequence anatomy in low, moderate and high subsidence settings. *Geologische Rundschau* **82**: 189–202

Kidwell, S. M. and Bosence, D. W. J. (1991) Taphonomy and time-averaging of marine shelly faunas. In Allison, P. A. and Briggs, D. E. G. (eds) *Taphonomy: releasing the data locked in the fossil record.* Plenum, New York, pp. 115–209

Kidwell, S. M., Fürsich, F. T. and Aigner, T. (1986) Conceptual framework for the analysis and classification of fossil concentrations. *Palaios* **1**: 228–38

Kidwell, S. M. and Holland, S. M. (1991) Field description of coarse bioclastic fabrics. *Palaios* **6**: 426–34

Racey, A. (1994) Palaeoenvironmental significance of larger foraminiferal biofabrics from the Middle Eocene Seeb Limestone Formation of Oman: implications for petroleum exploration. In Al-Husseini, M. L. (ed.) *GEO '94 The Middle East Petroleum Geosciences* **1**: 793–810

Radley, J. D., Gale, A. S. and Barker, M. (1998) Derived Jurassic fossils from the Vectis Formation (Lower Cretaceous) of the Isle of Wight, southern England. *Proceedings of the Geologists' Association* **109**: 81–91

Schäfer, K. (1969) Vergleichs-Schaubilder zur Bestimmung des Allochemgehalt bioklastischer Karbonatgesteine. *Neues Jahrbuch für Geologie und Paläontologie, Monatshefte*, pp. 173–84

Seilacher, A. (1963) Umlagerung und Rolltransport von Cephalopoden-Gehäusen. *Neues Jahrbuch für Geologie und Paläontologie, Monatshefte* pp. 593–615

Tedesco, L. P. and Wanless, H. R. (1991) Generation of sedimentary fabrics and facies by repetitive excavation and storm infilling of burrow networks, Holocene of South Florida and Caicos Platform, B.W.I. *Palaios* **6**: 326–43

CHAPTER 7

Fossils for the stratigrapher and structural geologist

There is no doubt that a major restriction on [sedimentological] research is the inadequacy of stratigraphic correlation.

E. Mutti

If you choose to ignore life, then stick to the Archaean.

This chapter discusses the applications of fossils to stratigraphy – without which geology doesn't exist – and it shows how fossils can be used in determining the structure of an area or section, a prelude to the stratigraphy.

7.1 Introduction to stratigraphic applications

Until outcrops, sections and wells are correlated by time lines, there is no way of gaining any real appreciation of the temporal distribution of past environments across an area or within adjacent basins and ranges, let alone of clarifying what was going on at distant points on the globe. Furthermore, until the rocks are dated, it is not possible to get far in understanding the structural development of the area.

Through much of the Phanerozoic, and in a general way also in the Proterozoic, fossils offer the most precise means of correlation available. There are several methods of correlation other than by using fossils, and Table 7.1 shows some of them with the service area for each, and the readiness with which each may be used in the field. Notice that many of the lithological methods are more distinctive than palaeontological methods, and that there are many methods that can be used for a small area. Also, fossils do not provide the only means of correlating on an international (chronostratigraphic) scale (Fig. 7.1). No mention is made in Table 7.1 of geochronology (absolute dating) using radiometric methods, because

it is not applicable for a field investigation and it is the least precise, though it does provide a framework for litho-, bio- and chronostratigraphy. So be on the lookout for rocks such as tuffs and those containing autochthonous fresh glauconite that might be radiometrically dated (to <1 m.y.).

Two major advances in stratigraphic procedure have been made in the past decade. The concept of *sequence stratigraphy* is now widely applied; the recognition of unconformities and their correlatives (in deeper water where the effects of emersion or shallowing are reduced), and *high-resolution sequence stratigraphy* (Fig. 7.4) for more local correlation (from one kilometre to tens of kilometres). Neither is fundamentally new and the high-resolution technique integrates all biostratigraphic data with other potentially correlative features, from marker beds (e.g. tuffs) to acmes and floods of body or trace fossils. Some, like shell beds, which formerly were often thought to be of correlative value, must be used with great caution. High-resolution sequence stratigraphy, using microfossil and trace fossil evidence, has been applied successfully to biosteering; controlling the drill bit for several kilometres parallel to the bedding (± horizontally). Sample analysis here has to be carried out speedily!

Allostratigraphy or sequence stratigraphy combines lithostratigraphy with chronostratigraphy to form a stratigraphy that is, in theory, applicable on a global scale. The global dimension is justified by invoking changes in sea level, brought about by glacial accretion and wastage, and crustal deformation as the principal mechanisms. The global or basin dimension to

Table 7.1 Some features of correlative value.

Feature	Field identification	Normal range (km²)	Comments
LITHOLOGICAL			
Turbidite event bed	poor to good	$<10^4$	Amalgamation may reduce correlative value;
Sheet sandstone	poor to moderate	$<10^2$	may be diachronous
Soil horizon	good	$<10^3$	
Thick coal seam	moderate	$<10^3$	Splitting may make identification difficult
Marine black mudstone	good	$<10^4$	Often excellent for widespread correlation when linked with biota
Shell beds, bone beds	distinctive	$<10^2$	Use with care, often unreliable
Hardgrounds	generally good	$<10^4$	May be local; useful if grouped into a pattern
Sequence boundary	generally good	10^4 to 10^6	Local to international
Tuff	good to very good	10^4	Especially useful if grouped
Iridium event[a]		$>10^6$	Should be worldwide, but needs specialist for identification
Milankovich cycles	poor to good	$>10^6$	Needs very careful analysis
FAUNAL AND FLORAL			
Plant, coral, brachiopod	poor to good	$<10^4$	Relatively easy to use with a suitable reference text for local biostratigraphic correlation
Ammonoid, graptolite, trilobite zonal taxa	poor to good	$>10^6$	Often of international application
Spore, foraminifera, coccolith zonal taxa[a]		$>10^6$	
Floods	distinctive		Determine autochthony; use with care
TRACE FOSSILS			
Vertebrate and trilobite traces	distinctive		Can be of international extent
Trace fossil flood	distinctive		Can be of basinal extent
GEOPHYSICAL			
Seismic reflector		$<10^4$	Generally of basinal application
Magnetic reversal		$>10^6$	Chronostratigraphic application
Magnetic susceptibility	can be good	$<10^6$	Best in siliciclastic sediments
Gamma-ray spectrography			
GEOCHEMICAL			
Isotopic peaks (e.g. ¹⁸O max.)		$>10^6$	Best in siliciclastic sediments

[a] Usually identified in a laboratory.

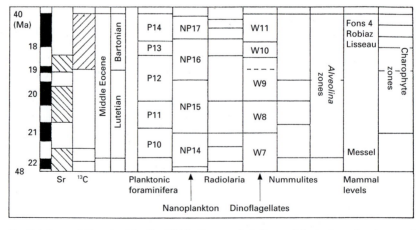

Fig. 7.1 A correlation chart for the Middle Eocene, Lutetian and Bartonian, showing zonal schemes for various groups of fossils against radiometric scale (Ma), magnetic anomalies 19–22, and chemizones marking changes in isotopic composition, strontium (Sr) and carbon (^{13}C). Adapted from Cavelier and Pomerol (1986).

Table 7.2 Degree of biostratigraphic refinement.

1000 Ma	(650 Ma)	Phanerozoic
100 Ma		System level: Devonian
10 Ma	(5–10 Ma)	Stage or series level: Palaeocene Series, Kimmeridgian Stage
1 Ma	(0.5–5 Ma)	Zone level: *Didymograptus bifidus* biozone

any study will be argued about in the laboratory or office, but it will be on the basis of observations made in the field. There, the key aspects are in recognizing and interpreting the facies successions and their bounding surfaces. The scheme can be applied, again in theory, to non-marine depositional environments if correlation can be made with climatic changes (due to sea-level change) that are registered in the facies. For further discussion see Miall (1997) and Reading (1996).

Every fossil has some stratigraphic value (Fig. 8.30), so the smallest sponge spicule or tiny lingulid brachiopod is indicative of the Phanerozoic rather than the Precambrian (and this may be important). It is useful to define the degree of stratigraphic refinement, required for the particular job in hand. There are approximately four orders of magnitude in stratigraphic refinement (Table 7.2). At system and series level, one is generally concerned with the range of a genus or higher taxonomic category. There are a number of older but well-illustrated texts that are useful at this level for identification, if it is appreciated that advances will have been made since their publication in definition, nomenclaure and, possibly, actual ranges cited.

Determination to zone level requires an appreciable amount of knowledge about the biota, and generally this is a job for the specialist. However, it is as well to know what is required. A really useful stratigraphic fossil needs five qualities:

- It should be fairly common.
- It must be widespread, ideally on an international scale.
- It should be present in many facies.
- Its stratigraphic range should be narrow.
- There must be good description (with illustration) of the species.

There are few organisms that can fulfil all these qualifications, and many monographs do not adequately describe the species or cover the palaeoecology. Nevertheless, try to find out about the biostratigraphy of the area it is intended to visit in order to get an appreciation of just how good the biostratigraphy is, and to get the most from the fossils that will be collected.

Collecting fossils can take time. It is important that the lithological succession is established in some detail, so the stratigraphic level of the fossils is known accurately. The primary objectives in mapping and well logging are to establish the stratigraphical succession and distribution of lithologies. This is followed by correlation of the stratigraphical units within and beyond the area under investigation. The duration of time represented by omission surfaces can be assessed from the nature of the contact and the biostratigraphy, the nature and derivation of intra- and extraclasts (including derived fossils, Box 6.2) and by tracing the omission surface into conformity.

7.2 Stratigraphic procedure

Fig. 7.2 attempts to show the relationships between the various stratigraphies. In the field or core shed, the problems in defining formational boundaries depend on the completeness of outcrop or core. Over the past 200 years (and more) an abundance of formation names of one sort or another have been introduced. Most are well established in the literature, but when inspected in detail can be seen to be based very loosely on lithological or fossil characters and with poorly or undefined formation boundaries.

Sequence stratigraphy has sought to put stratigraphy into a global framework. But, in practice, much of the geological column is made up of three types of succession, which do not exhibit the ideal and predictable response to sea-level change seen in a gently subsiding shelfal sedimentary environment:

- Thick successions of deep-water sediments, receiving sediment pulses unpredictably from various sources.
- Successions of varying thickness accumulating in local basins subject to intermittent tectonic activity.
- Condensed successions.

Unfortunately, attempting to carry sequence stratigraphy to areas with such successions has not always been fruitful. In the field two approaches can be used to define stratigraphic units.

Distinguishing boundaries In the three situations highlighted above, three types of boundary (Fig. 7.3) can be distinguished:

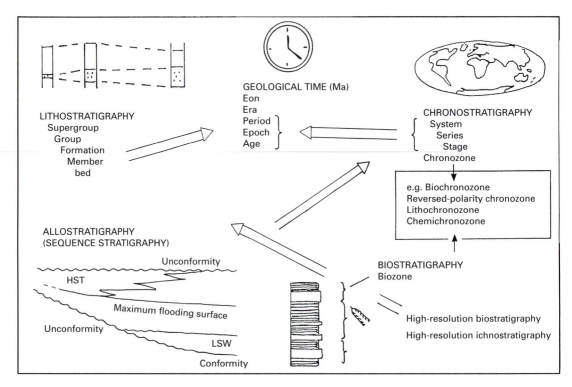

Fig. 7.2 Relationships between geological time, lithostratigraphy, allostratigraphy (sequence stratigraphy), biostratigraphy, high-resolution stratigraphy and chronostratigraphy.

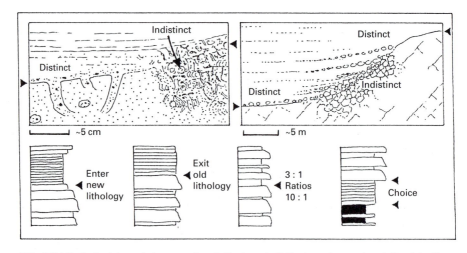

Fig. 7.3 Lithological boundaries can be distinct or indistinct, and may be defined in three main ways.

- The *appearance* of a new lithology, such as the introduction of a black, laminated shale into a sequence of muddy shales and sandstones.
- The *termination* of a lithology, such as the termination of graded greywackes in a sequence in which they are prominent, but with continuation of shales.
- The *gradual* change in lithological proportions, such as sandstone/shale ratio. There appears to be no significant event that led to a facies change, but when the succession is viewed overall, there is an apparent need to divide the succession, if only to make for more manageable units. An arbitrary decision will have to be made, e.g. on the sandstone/shale ratio.

There can be a conflict of interest on boundaries between the requirements of the map-maker and the interests of the sedimentologist or palaeontologist. In Fig. 7.3 (right) there is a choice of boundary, which may be placed either at the appearance of thick sandstones or at the disappearance of black shales. A boundary placed at the appearance of the thick sandstones is likely to be more easily mapped, but some might argue that the disappearance of black shales might be taken as a more significant event. In practice, defining boundaries on the criteria shown in Fig. 7.3 can often be more complicated.

Allostratigraphic framework Attempt to define an allostratigraphic framework to determine the stratigraphic units; try to appreciate the nature of events and try to recognize key surfaces (unconformity, and correlative conformity, flooding surfaces, maximum transgression surfaces) and the enclosed packets of sediments (parasequences). The palaeontological evidence is particularly important in defining the key surfaces (Fig. 8.28). In summary:

- Designate an accessible type section.
- Identify and describe the lower boundary precisely (the upper boundary is taken care of by the succeeding unit).
- Provide an illustrated description of the lithologies, including any lateral changes.
- Identify and describe diachronous facies.
- Identify and describe diagenetic facies (e.g. secondary dolomite).
- Determine the thickness and lateral extent.
- Determine the age.

Facies analysis is the most difficult part of a sedimentological analysis. In summary, one has to try to distinguish the various lithologies and group

them. The first stage is to log successions (Section 2.4). The bed is the basic unit (Fig. 2.2). It is characterized by its lithology, sedimentary structures, common fossils, and their vertical and lateral changes. Sometimes a bed can be unique, thus constituting a facies by right. Generally, beds can be grouped into facies. Many, if not most, sedimentary successions comprise either successions of facies, rhythms and cycles (Section 2.3), or event beds including turbidites, storm beds and ash falls (Sections 2.3 and 6.2), or combinations of them. Successions of facies (e.g. coarsening-up parasequences) reflect changing environments, and establishing these changes is the first stage of basin analysis. Formation boundaries will generally coincide with a particularly significant facies change. The old dictum that 'boundaries are drawn in the field and not in the pub' applies to formation boundaries and to facies boundaries. The only sure way is to walk them out. The distribution of fossils may be mapped independently of the distribution of lithologies, though this is done but rarely on a scale greater than about 1 : 10 000. Fossils closely associated with particular lithological facies are known as *facies fossils*.

Conflict can arise over the systematics of facies analysis. Recognizing and distinguishing fossil species and sedimentary facies are similar operations. The reputations of palaeontologists and sedimentologists depend on how well these operations are carried out. If we agree that a sedimentary facies represents the total make-up of a sediment, then problems can arise if the biological components change at a different rate to the hydraulically or chemically determined features. For instance, in a lagoonal succession (Fig. 7.5) there is repetition of four distinctive lithologies, each containing a rather sparse biota, but one that changes from dominantly freshwater at the base to near-marine at the top. It is generally better to recognize the four (litho)facies and to describe the change in biota independently.

Facies changes can be abrupt, gradational or by intercalation. Gradational changes are frequently due to bioturbation and were originally abrupt. Change in the proportion of an intercalating lithology, e.g. shales in sandstones, can be quantified.

At outcrop, and especially in core, the relief at formation bounding surfaces (parasequence and sequence boundaries) may not be evident. This makes it all the more imperative that surfaces are traced laterally, wherever possible, and that evidence for omission is assessed and evaluated. The interval represented by omission can be assessed from the

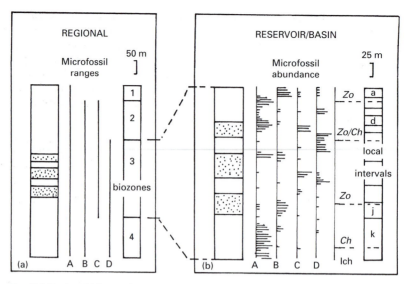

Fig. 7.4 Regional biozonation versus reservoir/basin high-resolution bio- and ichno-stratigraphy. (a) Biozones are plotted against disappearances (extinctions) and appearances (inceptions) of species. Sampling will have been at 20–50 m intervals and biozones in the order of 1–5 million years duration. (b) Potentially correlatable divisions plotted on the abundance of taxa and floods of the trace fossil (Ich) *Zoophycos* and/or *Chondrites*. Sampling will have been at 1–3 m intervals and the divisions of several hundred thousand years duration. Correlation may extend for 1–10 km and assist in defining parasequences. Other sections or wells can be correlated using graphic methods. Adapted from Simmons and Lowe (1996).

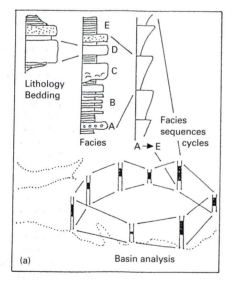

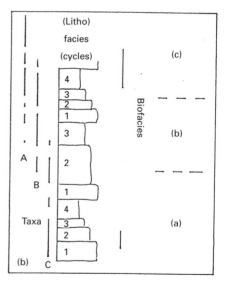

Fig. 7.5 (a) Basin analysis begins with the analysis of a bed; then follows facies recognition and analysis of successions of facies (parasequences) before basin analysis can be undertaken. (b) Relationships between lithofacies and biofacies (see text).

nature of the contact, the nature and provenance of intra- and extraclasts and derived fossils resting on the surface (Box 6.2), the ichnology of the surface (Fig. 8.28) and the biostratigraphy.

7.3 Biostratigraphy

How does one know which fossils will be useful in correlation? There is no simple answer to this question, and much will depend on what preparation has been done, and on the age of the strata (Table 7.3).

Biostratigraphy is the organization of sedimentary successions based on biological events (Figs 7.6 and 7.7): appearances (inceptions), disappearances (extinctions) and, less commonly, acmes of taxa (change in a ratio can also be used). Changes in the fossil assemblages through a more or less continuous succession can be attributed to four possible natural causes (providing it is possible to eliminate taphonomic effects, e.g. early shell dissolution and local dolomitization):

- Changes in ecological tolerance, e.g. substrate preference, salinity, temperature, turbulence).
- Migration, e.g. where changes in palaeogeography have allowed land animals to migrate, or marine organisms to increase their geographical distribution.
- Evolution.
- Extinction.

Table 7.3 Important groups of biostratigraphic fossils.

International	Regional	International	Regional
Precambrian		*Permian*	
stromatolites		ammonoids	miospores
acritarchs		foraminifera	vertebrates[c]
Cambrian			brachiopods
trilobites[a]	trilobites	*Triassic*	
acritarchs	archaeocyathids	ammonoids	vertebrates[c]
	acritarchs		spores
	Cruziana		vertebrate tracks[c]
Ordovician		*Jurassic*	
graptolites	trilobites	ammonites	ostracodes
conodonts	brachiopods		foraminifera
chitinozoans	acritarchs		dinoflagellates
	Cruziana	*Cretaceous*	
Devonian		ammonoids	calpionellids
graptolites[b]	brachiopods	coccoliths	ostracodes
ammonoids	corals	planktonic foraminifera	belemnites
conodonts	cricoconarids	dinoflagellates	radiolarians
ostracodes	trilobites	belemnites	Bivalvia[d]
	fish[c]	inoceramid bivalves	echinoids
	miospores		spores and pollen
			benthonic foraminifera
Dinantian		*Cenozoic*	
ammonoids	trilobites	coccoliths	diatoms
conodonts	corals	planktonic foraminifera	benthonic foraminifera
miospores	brachiopods	pollen and spores	ostracodes
foraminifera		dinoflagellates	bivalves
Silesian		radiolaria	gastropods
ammonoids	miospores		mammal teeth[c]
conodonts	macroplants		echinoids
foraminifera	non-marine		charophytes
	Bivalvia		
	ostracodes		

[a] Especially agnostids. [b] In Lower Devonian.
[c] In continental facies. [d] For example, *Inoceramus*, rudists in Tethys.

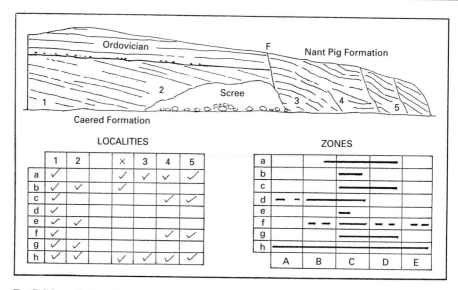

Fig. 7.6 A much simplified example of biostratigraphic correlation based on a Middle Cambrian sequence in North Wales. Eight fossil species (a–h) were collected at coastal sites (1–5), and at ×, an inland locality of uncertain stratigraphic position, but which is probably older than locality (3). A fault separates the Caered and Nant Pig Formations at the coast. The range of each species (a–h) elsewhere in Britain is shown right with the recognized zones (A–E). It would seem that the Caered Formation and the lower part of the Nant Pig Formation can be correlated with the lower part of zone (C), the main part of the Nant Pig Formation correlating with the upper part of zone (C) and probably ranging into zone (D). Highest point of cliff = 100 m.

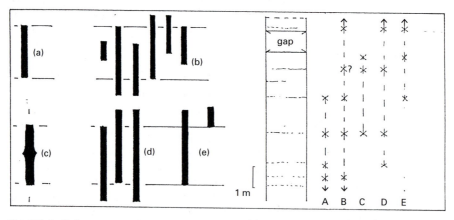

Fig. 7.7 (Left) Some types of biozone: (a) total range biozone, (b) assemblage biozone, (c) acme biozone, (d) concurrent range biozone, (e) interval biozone. (Right) A section from which fossils (A–E) are recorded at the points indicated. What types of biozone can be drawn? For example, a total range biozone can be drawn for fossil A.

With migration and changes in ecological tolerance, further ecological change may have resulted in the taxon (taxa) reappearing higher in the stratigraphy – the Lazarus effect. Evolution can also appear to be an extinction (pseudoextinction) when faunal or floral lists are examined, because of name change(s) at genus or higher level coinciding with a chrono-stratigraphic boundary (usually a system boundary). Extinction at genus or higher level must affect the whole clade. There are two other aspects that are at least as important, if not more important: preservational failure and collecting and sampling failure. Theoretically, one might argue that preservational failure must be related to the sedimentology of the unit, but the reasons may be elusive (Section 7.5.1). Collecting and sampling failure covers the 101 reasons why specimens were not recognized or collected.

Traditionally, sedimentary successions have been divided into zones (below). For the most accurate biostratigraphic correlation, only changes attributable to evolution and extinction can be used on an international (chronostratigraphic) scale. Although extinction can be reasonably inferred when taxa cannot be found above a certain bed (or level in a well), distinguishing evolutionary from ecological change or migration, when sampling through a succession, demands very careful analysis of the evidence. For much of the Phanerozoic, and for many regions, tables of zones are available, and these can be referred to. But consider that such schemes have been produced by careful analysis and judgement of data from many locations. One is not usually in a position to check the data or concur with the judgement. Neither is it generally possible to be in a position to check that the order of each event in each sequence is the same.

It is an axiom of biostratigraphy that where a succession of zones is the same at widely separated sections, then they are isochronous. If this were not so, then we should not be able to construct our palaeogeographies with much confidence. But it is always wise to consider the possibility of diachronism, i.e. that the bioevent did not occur everywhere at the same time. Also be ready to consider that dissimilarity of succession may be due to incomplete collection or to misidentification.

7.4 Procedure in biostratigraphy

Several operations are needed in order to reach a point from which to begin correlation of the section (or core) with other sections (or cores) and with established scales:

1 Collection of fossils and/or sampling of sediment for microfossils, with reference to the lithostratigraphy and allostratigraphy).
2 Preparation and identification (Appendix A and Chapter 4).
3 Keying the lithostratigraphy and allostratigraphy to the fossil data.
4 If the area is unlikely to be closely linked with well-established stratigraphical scales, then a local biostratigraphy must be generated. The object is to collect material that is going to be stratigraphically useful. Only rarely will more than a fraction

of the biota be of such use at zonal level. Attempt to locate this material, and check that specimens are sufficiently complete (e.g. trilobite cephala), adequately preserved, and free of compactional and tectonic distortion for full identification. But a fragment of a compressed Mesozoic ammonite may be very much more useful than any amount of most other fossils. Although in practice steps 2 to 4 are carried out in the laboratory, they are based on evidence collected in the field, and should be done with the participation of the field geologist.

5 Correlation with the regional zones and stages is generally the next operation (Fig. 7.5). For this it is necessary to have a good appreciation of what is meant by the zones and stages, and which groups of fossils are recognized as being most useful for correlation. At most periods of the Phanerozoic, more than one fossil group can be used, e.g. in the Upper Devonian, separate successions of zones can be recognized for ammonoids, trilobites and conodonts (see also Fig. 7.1). The boundaries of the zones for each group are seldom coincident. Tables of species ranges are useful. To compare the distributions of fossils in the area under investigation with the regional zonal scheme, the local distributions have to be tabulated. It will be surprising if the local ranges tie in exactly with the regional zonation. Often it is not possible to make more than a subjective judgement. Obviously discard from consideration long-ranging taxa, and see how species with narrower ranges fit.

It is useful to appreciate what the published schemes of zones mean, and there are several ways in which a zone, often designated by a single taxon, can be defined (Fig. 7.7). Zones are the most accurately known division of the biostratigraphic column, and by implication, of stratigraphic time. But zones, since they concern organisms, cannot have complete global distribution. Consider the distribution of living species today. Even the most widely distributed spores and pollen grains do not get everywhere! But a group of zones (a stage) can generally be used for correlation on an international scale:

• An *assemblage biozone* is a certain assemblage of taxa that is characteristic of a body of strata, but individual taxa are not confined to the particular interval of rock. This type of biozone is probably the most common type, and it may be designated by one taxon, often called the 'index' species. By bad luck this particular taxon may be absent in the section under consideration.

Box 7.1 The Channel Tunnel

Engineers for the Channel Tunnel (Fig. 7.8) had to design a route that follows closely the Chalk Marl (lower part of Lower Chalk, Cretaceous), which has the best tunnelling characteristics (low permeability, high strength and ease of penetration). The underlying Gault Clay had to be avoided because it is an expanding clay. The design also had to predict the effect of faults, many of which trend normal to the route, throwing the stratigraphy. The Chalk Marl is cyclic (0.5–1.0 m), but in the confines of a tunnel it is not possible to recognize the exact level on the opposite side of a fault. Recourse has to be made to the biostratigraphy and following sponge horizons. Changes in the planktonic and benthonic foraminifera, including abundances, coupled with local lithological information allow the stratigraphy to be determined to 1–2 m. Benthonic foraminifera can only be used locally, whereas the planktonic forms allow long-range correlation (Section 2.5.1). Precision to a few centimetres can be obtained by coring down to the Gault Clay.

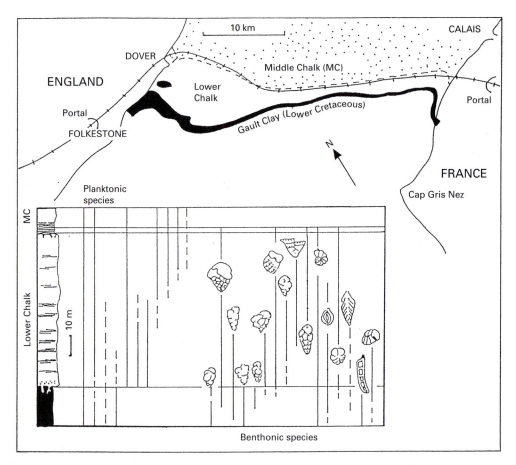

Fig. 7.8 Route of the Channel Tunnel, summary stratigraphic log and indication of ranges of selected planktonic and benthonic foraminifera determined at outcrop. Note the extinction event in the marls at the top of the Lower Chalk.

Sources: Destombes and Shepherd-Thorn (1971), Carter and Hart (1977)

- A *total range biozone* comprises the strata associated with a particular taxon.
- An *acme biozone* is characterized by an exceptional abundance of a taxon. You may well be suspicious. One thinks of possible ecological controls, but if the evidence shows an abundance, it should not be ignored on theoretical grounds.
- An *interval biozone* denotes the strata between two distinctive biostratigraphical horizons, e.g. the appearance of taxon A and, stratigraphically higher, the appearance of taxon B.

Another type of biozone is the *concurrent range biozone*, comprising strata characterized by overlapping ranges of specified taxa. The Lower Palaeozoic graptolite zones, as presently understood, are mainly of this type. Combinations of overlapping and total ranges can potentially offer greater stratigraphic resolution. In using many types of microfossil, lineages are applied. There is often a profusion of material, so it is possible to use ratios of species or abundance peaks. Occasionally, subzones will be encountered. Sometimes they refer to local subdivisions recognizable by the same criteria as zones, but caution should be used in extending subzones widely and assuming isochroneity. Horizons dominated by a single taxon should be correlated with care. In the Chalk (Upper Cretaceous) many such *floods* are of basin-wide extent.

7.4.1 Biostratigraphic boundaries

In describing zones, it has been almost implicit that it is possible to place a boundary between zones. In the field it may be difficult enough to decide where to place a formation boundary. Biostratigraphic boundaries are much more arbitrary. As a well is drilled and cuttings examined, the appearance of a new taxon actually marks its extinction (if the stratigraphy is normal). Disappearances, as drilling proceeds, are generally impossible to define because of contamination by younger fossils from the walls above. Hence in a well being drilled, where cavings are likely, position the upper boundary (marking extinction) immediately above the appearance of a taxon. At an outcrop, or completed core, position the base (species inception) immediately below the first occurrence. These are obviously arbitrary solutions and much depends on the sampling interval, and the time spent on collecting. Box 7.1 explains how biostratigraphy was used to design the route of the Channel Tunnel.

7.5 Problems

7.5.1 Preservational problems

If only the fossils were common and evenly distributed through a succession! If a formation or member appears to be devoid of body fossils, note that many trace fossils, e.g. *Cruziana* (Fig. 8.12) and vertebrate footprints, can be useful stratigraphically. In sandy sediments, shells may have been dissolved out at an early stage. Search for local better preservation or shells in concretions (Section 3.4).

7.5.2 Sedimentological problems

Dilution In thick 'expanded' sequences of deltaic sediments or turbidites, the stratigrapher may be asked to subdivide the sequence, which may encompass, say, one zone. If evolution will not oblige, then recourse must be made to pragmatic alternatives, with perhaps local application only (Table 7.1).

Condensation and stratigraphic thinning In contrast, there are successions that are relatively thin for the amount of time represented. The important point to appreciate in the field is that the sequence may be difficult to locate just because it is thin. Once an outcrop of such beds has been recognized, its nature can be determined. Stratigraphic thinning can arise in a number of ways: the area was starved of sediment (e.g. oceanic rise), or sediment did not accumulate, though sediment may have been temporarily deposited only to be reworked (bypassing). In sediment starvation, sampling may have to be at very close intervals. Condensation can occur by penecontemporaneous erosion or winnowing, leading to an accumulation of heavy shells and concretions. There may be significant stratigraphic mixing or even inversion of the residual fossils. Note carefully any differences in preservation mode and observe which stratigraphically important fossils also occur in the underlying strata, so as to spot fossils that were contemporary with the condensation event. Problems come when, for instance, three-dimensional phosphatized ammonites lying above an omission surface in the Cretaceous Gault Clay of south-east England, have to be compared with compressed ammonites in the clay below.

Table 7.4 Criteria for distinguishing between condensed and expanded successions.

Condensed	Expanded
Cycles, motifs often incomplete or poorly developed and may only be inferred from facies and intraclasts	Cycles, motifs complete and distinct (tidal, fluvial, deltaic cycles, etc.)

FACIES

Condensed	Expanded
Sharp and generally erosional contacts between facies, bioturbated	Gradational change between facies, necessitating arbitary boundaries
Event beds poorly reprsented or amalgamated	Event (storm) beds with complete (hydraulic) successions of sedimentary structures, but amalgamated in high-energy environments
Thin bedding typical, cross-bedding poorly developed and represented by lags and toe sets; channelized facies overrepresented	Bedding poor to good; cross-bedding often well preserved; channelized facies seldom dominant
Omission surfaces common	Omission surfaces uncommon
Mud layers underrepresented, but mud associated with intraclasts and bioturbation structures	Mud layers typical
Coal and lignites poorly represented, but root beds prominent	Thick coals (±)
Pedogenic structures common but thin	Pedogenic structures often associated with coals

SHELLY BIOTA

Condensed	Expanded
High corrasion, fragmentation, disarticulation, dissociation	Low corrasion, fragmentation, disarticulation, dissociation
Shells uncrushed, filled	Shells crushed ± unfilled ± dissolved
Major hydraulic concentrations	Minor hydraulic concentrations
Major transgressive shell beds	Transgressive shell beds generally absent
Hard substrate shells generally overrepresented	Infaunal shells common
Shelly fauna overrepresented	Shelly fauna diluted
Nekton and plankton underrepresented because of taphonomic considerations	Nekton and plankton diluted
Reworked, concretionary fossils common (e.g. phosphatized shells)	Few reworked shells
Autochthonous shells rare and only deeper burrowers represented	Authochthonous shells relatively common
Higher diversity of opportunists and specialists	Opportunists associated with event beds; specialists underrepresented

BIOTURBATION AND ICHNOFABRIC

Condensed	Expanded
Bioturbation grade high	Bioturbation grade variable, but often confined to tops of event beds or to particular intervals of cycles
Complex and often indistinct ichnofabrics (±)	Ichnotaxa generally distinctive and readily recognizable
Tiering complex because of overprinting	Tiering generally distinct, especially in event beds
Soles often obscured because of multiplicity of traces, leading to poor bedding definition	Sole traces distinct
'*Ophiomorpha*' rarely pellet-lined	In sandy facies *Ophiomorpha* typically with pellet-lined margin
Escape structures uncommon	Escape structures common

Extent of condensation It is important in basin analysis to try to evaluate the extent to which successions are condensed or expanded relative to each other, or whether it is only parts that are relatively condensed, indicating fluctuation in rate of subsidence. Pelagic condensed successions can be readily identified by their closely spaced zones, and thick sussessions of turbidites are, virtually by definition, expanded. The problem is often regarded as simply one of accommodation.

Neptunian dykes Clastic (Neptunian) dykes represent the filling of cracks that opened on the seabed. One or more events may have occurred, and the crack may not have reopened symmetrically. Neptunian dykes can be large or very thin and may pass into bedding-parallel 'sills'. Younger material piped down (stratigraphic leakage) can be difficult to detect. Map out such structures carefully.

Leakage and mixing Leakage and mixing can also occur when microfossils are washed into burrows, or reworked by bioturbation either downwards or upwards. In both, stratigraphic definition will be reduced.

Reworking of fossils It is surprising how readily small resistant fossils may be reworked into younger or older sediments, e.g. Coal Measure (Carboniferous) spores into Tertiary sediments in southem England, or Pleistocene foraminifera into Eocene clays in Timor. With macrofossils, watch for specimens which have a mode of preservation decidedly different from the associated biota, e.g. phosphatic internal moulds (Box 6.2, Fig. 8.30).

Boulder beds and slumps Intraformational boulder beds and slumps may require careful inspection and sampling. The boulders can be younger or older than the matrix.

Criteria in Table 7.4 Table 7.4 lists criteria that may be considered for discussion when condensed or expanded successions are suspected; the criteria are for siliciclastic nearshore and deltaic facies.

7.5.3 Ecological problems

There are three types of ecological problem. First, facies fossils are generally recognizable as such by their association with a particular lithology. Some may be long-ranging species that follow the lateral and diachronous facies shift, or turn up each time a facies reasserts itself (as with many bivalve molluscs), but others may exhibit a degree of morphological change, or a different species may appear. Such changes may be useful for local correlation, or even for correlation into another basin (e.g. with many brachiopods). Second, even the most widely distributed fossils are to some extent restricted in their distribution, e.g. Jurassic ammonites are virtually unknown in reefal facies. Third, a local facies change may coincide with a disappearance, but this may not be an extinction.

7.5.4 Bibliographical problems

Stratigraphers and palaeontologists, perhaps more so than other geologists, have to use and indeed rely on older literature. The expert will be able to make a shrewd, modern assessment of older descriptions and what an older taxon name is likely to signify now. The stratigraphy was often done in considerable detail, though not often with good facies analysis, and often assuming greater lateral extent of quite thin beds than would be considered reliable today.

7.5.5 Resolution problems

There is a limit to the fineness of the divisions into which any evolutionary lineage can be separated. Whereas it is true that for some parts of the stratigraphic column, certain groups, e.g. graptolites and ammonites, evolved so quickly that zones of about 0.5 Ma can be recognized, palynological zones of the Jurassic have an average duration of about 4 Ma.

7.6 The graphical method of correlation

The graphical method (Shaw, 1964; Mann and Lane, 1995) of correlating sections is simple to use even in the field, and indeed, should follow construction of cross-sections. A few sheets of graph paper are needed. The method involves plotting graphically (Fig. 7.9) the information obtained from two or more sections. The linear relationship, 'regression line', will indicate the best correlation and relative sedimentation rates between the two sections. Any lithological event recorded (tuff, shell bed, storm bed, etc.), as well as biological data, can be plotted. For biological data, plot appearances, disappearances and frequencies. In Fig. 7.9 (a) the majority of appearances and disappearances cluster about the line $x = y$, but it is readily seen that the ranges of taxa are not coincident. This can be due to four possibilities: (1) incomplete

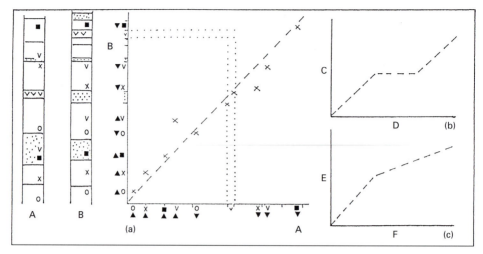

Fig. 7.9 Graphical (Shaw) method of correlation. (a) Data from section A is plotted against data from section B: (▲, ▼), respectively, indicate appearances and disappearances of taxa (■, ×, ○, ∨). (b, c) See text.

sampling, (2) incomplete exposure, (3) actual differences between the two sections due to ecological factors, or (4) actual differences due to geographical factors. The distribution of lithologies is similar in each section, though the tuff horizons (line of ∨ symbols) are not equivalent (as might be considered on an initial reading). While in the field, it would be sensible to search further for species X in section B, to determine whether higher occurrences may be present.

In Fig. 7.9 (b) the offset in the correlation line means a sedimentation break, or a fault cuts out part of the sequence in C. The evidence must be in the field! In Fig. 7.9 (c) there is a marked decrease in sedimentation rate indicated in E relative to F for the upper part of the section. What lithological evidence would be expected?

If two sections correlate well, then sedimentation rates in both are uniform, but not necessarily constant, since one is not correlating actual time. Do a few exercises to help understand the method, by increasing sedimentation rate in one section, and introducing a thrust fault, or slump (with or without erosion of the sole). This method dispenses with zones.

Another 'non-zonal' method of correlation is to compare statistically (e.g. cluster analysis) the types and frequencies of fossils from each sampling point.

When relaxing one evening, ponder over the questions: Why can we not dispense with zones? And, might a zonal scheme still be acceptable if it contains unassigned gaps, representing strata from which no zonal fossils have been obtained?

7.7 Fossils in structural geology

There are several areas where structural geologists and geophysicists can make use of fossils in the field. In many cases the palaeontological evidence provides an independent alternative to structural and geophysical evidence. But the palaeontological evidence may be no more exact than the geophysical evidence, because of lack of documentation (the species has yet to be described), or because of the inherent lack of precision of palaeoecological data (many organisms are tolerant of a wide range of environments). Also, relatively few species have a short time range.

7.7.1 Applications to biostratigraphy

Evidence from fossils is needed to determine stratigraphic succession, time gap at unconformities and disconformities, rates of sedimentation (Section 2.5), and as a check on isotopic age. In a structurally complex region, relative ages are needed particularly for structurally isolated units. The approach here is rather different from that required by the stratigrapher for fine correlation. In general, the need is to determine, by any means possible, the age of particular packets of rock. Often it is adequate to assign the age to stage or series level (Section 7.1), rather than to the more precise biozone.

Macrofossils may be useful, but distortion and/or recrystallization may render determination difficult or

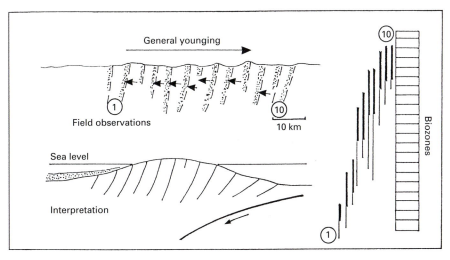

Fig. 7.10 Application of biostratigraphy in establishing nature of an accretionary prism and (right) time stratigraphy of fault slices 1–10. Thicker line segments refer to turbidites, thinner parts to basinal black shales and chert. Adapted from Leggett *et al.* (1979).

inconclusive. Carbonization of organic matter or replacement by a micaceous mineral is common. Micaceous mineralization is reflective. When examining a suspect mudrock, rotate it to try to spot such preservation. Fossils are generally disposed parallel to bedding, so it pays to locate bedding and to try to break the rock along it. The rock probably tends to break along a cleavage. Some weathering of the rock is usually required for success. Since they accommodate to deformation without significant distortion, small fossils such as ostracodes can be particularly useful, and may be found on the edges of cleavage plates. For instance, fingerprint ostracodes are invaluable in the late Devonian and Carboniferous thrust and nappe belt of south-west England. Use a hand lens on reduction spots: there may be a conodont at the centre! Calcareous sediments, even if somewhat impure, or calcareous lenses between lava pillows will be worthwhile collecting for later acid digestion to retrieve conodonts (Ordovician to Triassic).

Dating the rocks either side of a major fault at a number of points along the fault will indicate the time, or times, and direction of movement. Anticipate more than one movement, and in different directions. Fossils in boulders and olistoliths will help date their emplacement time. But like any other sedimentary particle, fossils can be reworked; and microfossils can be reworked with surprisingly little damage. Sample each type of clast and also the matrix, since the clasts may be either younger or older than the matrix.

An appreciation of the rate of uplift for geologically young shallow-marine sediments can be obtained when they are found at considerable elevation (e.g. in the Appenines), or at depth (submerged reefs), if they can be dated biostratigraphically. The timing of nappe movement may be determined by the progressive younging of the overthrust sediment, caught between faults or in synclinal structures. This is likely to be associated with a progressive younging of the post-tectonic sediment, as with the onset of continental sedimentation from north to south in the Norwegian Caledonides. Although the general configuration of an accretionary prism that is still under construction today may be evident from seismic survey, fine biostratigraphy can be used to support or reject a structural hypothesis for an accretionary prism in older rocks. The general younging direction will be towards the site of the former trench and subduction zone, but with each fault-bounded section showing younging in the opposite direction (Fig. 7.10).

7.7.2 Application to facies analysis

Having determined the stratigraphic succession, the next stage is to find out about the sedimentary environments. The lithology may give sufficient definition, but the biota can be useful, e.g. to decide whether particular turbidites were deposited above or below carbonate compensation depth.

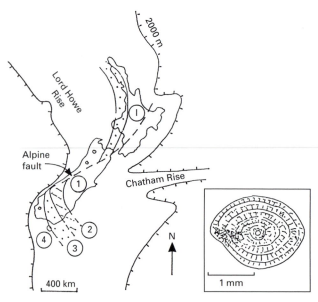

Fig. 7.11 New Zealand and adjacent platform, tectonostratigraphic terranes: (1) Torlesse, (2) Caples, (3) Hokuni, (4) Tuhua. Inset: fusuline in axial section. Adapted from McKinnon (1983).

Where structurally separated blocks differ in facies but not in age, it may be possible to estimate original separation distances from palaeoecological analysis. This applies particularly to thrust terranes where shore facies might be brought against deep-marine facies. Deformation may be too great to make detailed facies analysis, but try to use broad facies, such as shelf, outer-shelf, pelagic facies, as a start. Fossiliferous horizons often have considerable lateral extension, making them useful markers, but evaluate the context to make sure that only a single rock unit is involved.

7.7.3 Biogeographical applications

The recognition of tectonostratigraphic (suspect or displaced) terranes, blocks of crustal rocks accreted onto larger plates, follows naturally from plate tectonic theory. Here is an area where palaeomagnetic and palaeobiogeographic data can be combined. In most examples it is juxtaposition of fossils, which must have been separated originally by many degrees of latitude, that provides the clue to the magnitude of the displacement. The examples from the Pacific margins are now classic. For instance, New Zealand comprises at least four tectonostratigraphic terranes (Fig. 7.11) that had accreted onto Gondwanaland by the mid Cretaceous, prior to the area splitting off and moving northwards at the end of the Cretaceous.

Unit 3 contains late Palaeozoic and early Mesozoic faunas, mainly brachiopods, indicating close links with eastern Australia. But unit 1 contains Permian fusulines (spindle-shaped macroscopic foraminifera) generally associated with lower latitudes and suggesting a 5–10° southward movement by transform faulting of an independent block in post-Permian times (McKinnon, 1983). Fossils associated with shallow-marine carbonates are most useful because such sediments are most typical of low latitudes. Longitudinal displacement is more difficult to prove. Similarities between tectonically separated biotas can also be used as evidence for only minor original separation. Much depends on understanding the methods of faunal and floral migration.

7.7.4 Geopetal applications

Using body and trace fossils as 'way-up' (geopetal) criteria is well known. Unfortunately, since many strongly deformed sediments are pelagic, where grading may be none too obvious and sole marks rare, then any reliable criteria will be useful. Way-up criteria are also required in slumped but relatively undeformed sediments. The only biological criteria that are wholly reliable are U-burrows (Figs 8.18 and 8.19) – burrows normal to bedding – along with various biogenic sole structures (which never occur on bed tops), predepositional sole traces, and also tiered bed sequences (parallel to bioturbated; see Fig. 8.19). Many trace fossils are much alike on soles and tops, and it would be generally unwise to spend time trying to unravel the situation at deformed sites. The asymmetrical fill of *Zoophycos* (Figs 8.4 and 8.13) should be used with care, because a reversed asymmetry can occur near to the axial portion of the burrow system. When present, *Zoophycos* is usually common.

Using shell convexity needs care unless there is a clear geopetal indication below the umbrella of the fossil. The attitude of shells reflects the nature of the hydraulic processes (Section 6.4). Unfortunately, shells are often so flattened as to make them unreliable as way-up criteria. Corals and other organisms with a conical growth form can be used if they are positively in position of life.

7.7.5 Organic matter and palaeotemperature

All organic material (non-skeletal) undergoes thermal alteration at temperatures in the range 50–400 °C. The colour attained reflects the temperature and duration at which the temperature is maintained. Colour change is irreversible, but weathering may lighten the objects. The material is slowly carbonized with loss of hydrogen and oxygen. The end result is carbon – black and highly reflective. But different organisms, because of their differing compositions, exhibit a somewhat varied succession of colour changes associated with increase in reflectivity. Most attention has been given to spores, pollen and plant fragments in assessing hydrocarbon thermal maturation, but acritarchs, dinoflagellates, chitinozoans, graptolites, insects and fragments of crustaceans and cheilicerates can also be used. Obviously only the larger material will be spotted in the field, and it will be necessary to collect for spores (also useful for biostratigraphy). Organic-walled microfossils can be found in most sediments where the organic material has not been completely oxidized. Collect from fresh, darker sediments. Fossil charcoal (fusain) is known from the Devonian and is unchanged with temperature rise.

Conodonts, vertebrate teeth and bones incorporate organic matter within the phosphate, and also exhibit a range of colour change with increasing temperature. White conodonts in shales can be readily spotted, with a hand lens, indicating temperatures above 350 °C.

7.7.6 Fossils and strain

Deformed fossils, because they deform like the matrix, can often be used as a means of accurately determining strain in deformed rocks, either when the section is devoid of structures, such as pebbles, boudins and folds, or in an independent test. Search for trilobites or brachiopods that still retain symmetry, but have been stretched or squashed without shear. The trend of the principal axis of the strain ellipsoid can be measured directly. Use can be made of a group of deformed fossils with various orientations, especially when they had original bilateral symmetry (Fig. 3.24). The Wellman (1962) method is useful in the field to determine the strain ellipsoid. Linear fossils, such as belemnites and crinoid stems, are often boudinaged. Search for specimens that are still straight (without zigzag), and are thus parallel to the principal strain axis. Measure the stretched length (L_1), and sum the lengths of each original segment, which generally equals the original length (L_0). The elongation is thus $(L_1 - L_0)/L_0$. Sketch and photograph specimens showing zigzag.

Deformed vertical burrows (generally referred to *Skolithos* or *Diplocraterion*; Fig. 8.18 shows them undeformed) have frequently been used in strain analysis. Make measurements on bedding surfaces of the long and short axes of each burrow (from which the axial ratio can be calculated), the angle to the reference direction, and the plunge of the burrow axes. Make sketches and take photographs of marked surfaces normal and parallel to bedding. Include a scale. The burrows may have originally varied somewhat from being orthogonal to bedding. When collecting any specimen for later structural analysis, make sure it is clearly marked with dip, strike and way-up, and make sure the necessary measurements are recorded: dip, strike (or dip direction), cleavage and cleavage–bedding intersection, lineation. Small fossils, crinoid ossicles and pyrite framboids often display pressure shadows normal to the principal strain.

References

Carter, D. J. and Hart, M. P. (1977) Aspects of mid-Cretaceous stratigraphic micropalaeontology. *Bulletin of the British Museum (Natural History – Geology)* **29**: 1–135

Cavalier, C. and Pomerol, C. (1986) Stratigraphy of the Paleogene. *Bulletin de la Société Géologique de France* **8**: 255–65

Destombes, J. P. and Shepherd-Thorn, E. R. (1971) *Geological results of the Channel Tunnel site investigations 1964–65*. Report 71/11, Instiute of Geological Science, Keyworth, Notts

Leggett, J. K., McKerrow, W. S. and Eales, M. H. (1979) The Southern Uplands of Scotland: a Lower Palaeozoic accretionary prism. *Journal of the Geological Society, London* **136**: 755–70

McKinnon, T. C. (1983) Origin of the Torlesse terraine and coeval rocks, South Island, New Zealand. *Bulletin of the Geological Society of America* **94**: 967–85

Mann, K. O. and Lane, H. R. (1995) *Graphic correlation.* Special Publication 53, Society of Sedimentary Geology, Lawrence OK

Miall, A. D. (1997) *The geology of stratigraphic sequences.* Springer, Berlin

Reading, H. G. (ed.) (1996) *Sedimentary environments: processes, facies and stratigraphy.* Blackwell Science, Oxford

Shaw, A. B. (1964) *Time in stratigraphy.* McGraw-Hill, New York (Referred to in many texts on principles of stratigraphy)

Simmons, M. and Lowe, S. (1996) The future for palaeontology? An industrial perspective. *Geoscientist* **6**: 14–16

Wellman, H. W. (1962) A graphical method for analysing fossil distortion caused by tectonic deformation *Geological Magazine* **99**: 348–52 (Referred to in many texts on structural geology)

Further reading

Texts on fossils in structural geology

McClay, K. R. (1988) *The mapping of geological structures.* J. Wiley, Chichester

Park, R. G. (1997) *Fundamentals of structural geology*, 3rd edn. Chapman & Hall, London

Two standard guides to stratigraphic procedure

Hedberg, H. D. (1976) *International stratigraphic guide: a guide to stratigraphic classification, terminology and procedure.* J. Wiley, Chichester

Rawson, P. F. *et al.* (in press) *A guide to stratigraphic procedure.* Geological Society Special Report, Geological Society, London

A chart of international chronostratigraphic divisons

Haq, B. U. and Eysinga, F. W. B. van (1998) *Geological time scale*, 5th edn. Elsevier, Amsterdam

CHAPTER 8   Trace fossils and bioturbation

Trace fossils, first of all, are sedimentary structures.

Trace fossils and bioturbation record the activity of organisms and plants on and within the sediment (substrate) in order to maintain their physiological functions. They represent fossil behaviour. In unconsolidated sediment animals form burrows, tracks and trails, and plant roots penetrate the soil. In consolidated (lithified) sediment animal activity is recorded in borings, etchings and raspings. Traces and trace fossils are thus organic sedimentary structures. For convenience, fossil dung (coprolites), faecal pellets and sometimes fossil eggshells are included. Trace fossils, with a couple of exceptions, were formed exactly where they are now found, and represent the

responses to the physical and chemical nature of the substrate. Hence they are sensitive to changes in the environment. Organisms dragged over or impinging on the substrate leave tool marks (Section 6.6); the organisms were generally already dead.

The aim of this chapter is to show how trace fossils and bioturbation can be recognized in the field, and how their information potential can be realized. The applications of trace fossils in geology (Table 8.1) are only beginning to be appreciated. Much basic research remains to be done. Trace fossils have a wider significance than body fossils, but their significance is in many respects complementary. It is the

Table 8.1 Information potential of trace fossils and bioturbation.

Sedimentological	*Structural*
Indication of nature of substrate	Geopetal attitude and burrow fills
Omission surfaces and key stratal surfaces	Strain gauges
Grain size distribution, sorting and permeability	
Porosity distribution	*Facies/environmental interpretation*
Event sedimentation (tempestite, turbidite, etc.)	Broad and specific facies interpretation
Homogenization trend	Palaeobathymetry
Sedimentation rates	
Current indicators	*Palaeobiological*
Penecontemporaneous erosion	Evolution of behaviour
Skeletal degradation	Tiering and recolonization
Pseudolamination	Information on soft-bodied biota
	Appearance and evolution of metazoan life
Geochemical	
Oxygenation levels	*Geotechnical*
Salinity	Site location
Diagenetic signatures	
Stratigraphic	
Biostratigraphy (*Cruziana*, vertebrate footprint)	
Biosteering	
Biostratigraphic redistribution	

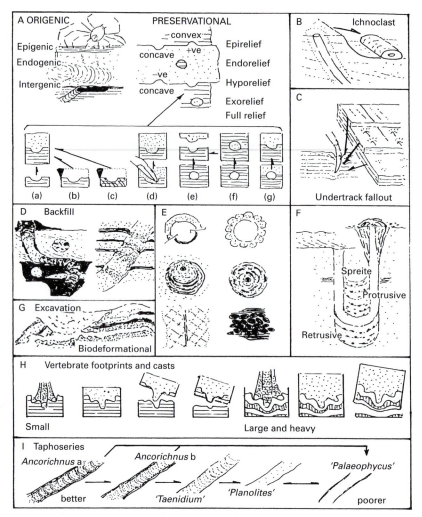

Fig. 8.1 Modes of trace fossil formation, preservation and terminology following Seilacher (1964) with modifications by Chamberlain (1971) and Jensen (1997). (A) General *origenic* (relation to sediment at time of trace formation) and *preservational* classification with seven possible interpretations of a convex (positive) hyporelief (hypichnion): (a) epigenically formed groove (above substrate); (b) groove made on an eroded firm substrate; (c) groove cut, or boring dissolved, in lithified sediment; (d) interface undertrack groove; (e) open burrow in mud eroded to expose groove; (f) filled burrow, in sediment subsequently eroded and groove exhumed; (g) burrow formed at sand–mud interface, with collapse of sand. (B) An *ichnoclast*, deeply impregnated (mucus) margins of burrow (e.g. *Solimyatuba*), penecontemporaneously eroded (see Fig. 8.30). (C) Undertrack fallout (see Fig. 8.2). (D) Manner of backfill of animal passing through heterolithic sediment. Backfill reflects lithology passed through. Burrowing at an interface results in mud flakes in sand. Backfill may be scarcely discernible when passing through a relatively thick sand or mud. Mud may become drawn into margin of a sandy fill. (E) Seven types of burrow margin, clockwise from top left: thick sandy or thin mud lining, respectively, *Palaeophycus heberti* and *P. tubularis*; pellet lining of *Ophiomorpha nodosa* (see Fig. 8.4); infiltrated mud, pushed aside; pellet-fill; diagenetic (siliceous) concretionary margin to *Bathichnus* (many times burrow diameter); passive fill in abandoned burrow with dual openings. (F) U-burrow (*Diplocraterion*) with spreite, and opening with 'pushed-aside' sediment (see Fig. 8.19). (G) A *biodeformational* structure made by fish (e.g. ray). (H) Vertebrate footprint formation and preservation: (left set) penetration of laminated sediment without apparent disturbance of lamination, fissile sediment may split at various levels to display less than complete cast; (right set) made by heavy animal in heterolithic sediment deforming layers (up to 0.5 m), sampling of lower layers shows a *transmitted (under)cast*. (I) Quality of preservation that can influence naming of a trace fossil illustrated by a meniscate and lined burrow; it is often possible to trace a poorly preserved burrow into better preservation, thus obviating misidentification; at worst preservation, *Ancorichnus* may resemble *Palaeophycus* or *Planolites*. Adapted from MacNaughton and Pickerill (1995).

sedimentological questions that need to be especially addressed, particularly the role of trace fossils in facies interpretation, for which a new approach is used, placing emphasis on colonization and tiering patterns. Naming trace fossils is discussed in Section 4.2, and general field strategy in Sections 2.8 and 2.9.

Fig. 8.2 Undertrack preservation of limulid track viewed from lower surface (Parrsboro Formation, Nova Scotia, Upper Carboniferous). The laminated siltstone has split along two levels, revealing a lower level with traces of anterior appendages, and a slightly higher level with additional traces due to the pushers and telson. The telson trace cuts a set of traces made from a shallow depth and below the substrate surface used by the first animal. Scale bar = 10 mm.

8.1 Principles

Sedimentological

1 Trace fossil are autochthonous. There are two exceptions: both are *ichnoclasts* (Figs 8.1 and 8.30, p. 163).
2 Trace fossils are modified during burial.
3 Trace fossils generally exhibit more information than the biological trace, e.g. preservation of *spreite*.
4 Different traces have different preservation potential (Fig. 8.2).
5 Rate of sedimentation influences burrow form (e.g. escape burrows) and degree of bioturbation.

Ethological (behavioural)

1 Trace makers are rarely preserved with their traces. And a body fossil found with a trace may not have produced it. It may represent a 'lodger' or nestler, or it may have been washed in.
2 Different types of animal can make identical traces. For instance, vertical shafts are today made by a variety of worms, arthropods and coelenterates.
3 The same trace may be differently preserved in different substrates.
4 A trace may change in character laterally, reflecting environmental change: *compound* ichnotaxa (Fig. 8.3). For instance, the pellet-lined burrow

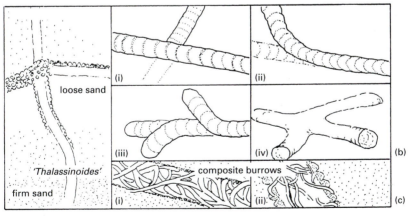

Fig. 8.3 (a) A compound burrow, *Ophiomorpha nodosa*, where the unlined shaft passes down to 'true' *O. nodosa* (but the right gallery has only a lined roof), then downwards to an unlined shaft where the trace penetrates firm sand. (b) Four modes of branching in trace fossils: (i) false branching is commonly seen where one burrow is incompletely preserved; (ii) secondary successive branching is where one burrow follows the course of another; (iii) primary successive branching is attributable to successive probes and withdrawals; (iv) simultaneous branching forms a network of open passages. (c) Composite burrows: (i) *Chondrites* pervading the fill of a large diameter arthropod burrow; (ii) filled *O. nodosa* penetrated by *Planolites*. Parts (b) and (c) adapted from Bromley (1996).

(*Ophiomorpha*) may pass into an unlined burrow (*Thalassinoides*), where the sediment may be inferred as having been firm and not requiring a supportive lining. In practice it is not helpful to refer to short sections by a different name, but any change should be described.

5 Several animals may be involved with a *composite* trace (Fig. 8.3). Filled burrows may be re-excavated (as a ready source of nutrients). Each trace-maker may leave evidence of occupancy.

6 The trace will be modfied by growth of the animal responsible, or the form modified by small fluctuations in many ecological factors.

8.2   Formation and preservation

Fig. 8.1 illustrates *origenic* mode, the ways in which traces are formed, and *toponomic* mode, how they have been preserved with respect to the substrate surface, substrate, or interface between contrasting lithologies (perhaps the most common mode at outcrop). It is important to appreciate whether the trace was formed before a (predepositional) sedimentation event, during sediment aggradation (syndepositional), e.g. some *Diplocraterion*, or following (post-depositional) a sedimentation event (storm, tempestite or turbidite). With *firmgrounds* and *hardgrounds* (Section 8.5.2)

the situation is more complex: the *preomission* suite represents softground colonization. The early stages of the *omission* suite carry traces that were able to cope with gradual lithification, and then (by boring and, of course, encrustation) the lithified substrate. Renewed sedimentation led to a *post-omission* suite of loose or soft substrate burrows that had to cope with a *concealed* (hidden) but irregularly perforated sublayer (Section 8.5.2).

Some workers contend that *Cruziana* (Figs 8.11 and 8.12, pp. 143 and 144), attributable to burrowing trilobites in the Palaeozoic, was formed epigenically on a muddy substrate and then a cover of sand cast the structure. But sectioning clearly shows (Fig. 8.11) that in most cases the sand cast represents complete disturbance within a sand layer and that the preservation is due to the 'chance' occurrence of a mud layer just below the substrate surface, a *concealed* change that 'fooled' the animal. The trilobite was actively exploring and sifting a sandy substrate. Other occurrences can be explained by washing-out and renewed sedimentation. Epigenic traces can generally be recognized by faecal castings (rare), 'levees' to a groove, and emergent burrows identified by funnel apertures or burrow restriction. Movement on and within the sediment may simply disturb the lamination but not produce a trace that has the potential to become a definite trace fossil (Fig. 8.1(G)). This can be seen when animals push or dig their way into laminated

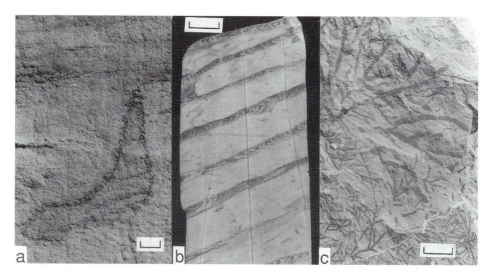

Fig. 8.4 (a) *Ophiomorpha nodosa*, in acetate peel, from sand at the locality depicted in Fig. 5.12. (b) *Zoophycos* in longidudinal section (White Rock, Wairdridge, Amuri Limestone, Eocene, New Zealand); with care, asymmetry of the laminae may be used as a geopetal indicator. (c) *Chondrites*, two species displaying colour differences with sediment (near Biarritz, southern France, Upper Cretaceous Flysch). Scale bars = 10 mm.

sandy sediment, or when the sand collapses below a tun-nelling organism. These *biodeformational structures* (Fig. 8.1(G)) and *collapse structures* can leave distinctive fabrics. *Excavation* structures, often large but of variable size and shape, are made today by fish and arthropods.

Both *undertracks* and *undertrack fallout* (Figs 8.1(C) and 8.2) result from the appendages of arthropods, and feet of light vertebrates, *penetrating* laminated sediment. Since differently structured appendages penetrate to different depths, the rock may subsequently split to show these differences.

True *footprints* or *footcasts* (Fig. 8.1(H)) represent the actual surface impacted by a foot. The best indication that one has a true footprint (or cast) is to find some evidence for skin impression. *Underprints* are formed by vertebrate feet that have excavated a laminated substrate, which was then partly fused to the sediment, burying the true *footcast*.

On soft to firm heterolithic substrates the foot of a heavy animal will deform the layers, leading to a *transmitted footprint*, or more commonly a *transmitted footcast*. The succession of casts will gradually fade out with depth. In more homogeneous sandy sediment the effect of trampling (dinoturbation) may lead to a high degree of deformation. The only way to distinguish this from soft sediment deformation caused by inorganic physical processes (e.g. load casting or slumping) is to show that the fabric is not due to physical processes – by no means an easy matter. Box 8.1 considers the Upper Jurassic rocks of the Dorset coast.

Most occurrences are *undertracks, underprints* (or *undercasts*) and *transmitted prints* (or *casts*). It is necessary to distinguish between *incision, excavation, transmission*, degrees of *sediment adhesion* and these in combination, together with *deformation* to the margin of a larger foot impact.

8.2.1 Burial and diagenesis

Trace fossil taphonomy may be considered in exactly the same way as the taphonomy of body fossils: formation, stratinomy and diagenesis (Fig. 3.1). As sediment accretion takes place, so deeper burrowers move through sediment already mixed by shallow bioturbators (Fig. 8.16). Unless the substrate surface, the *mixed zone*, can be preserved by burial (e.g. distal turbidite or deep-water tephra layer) or an ecological change halting activity, there is no real chance of its preservation.

In the *transition zone* compaction is the most important factor in trace loss. Deeper burrows penetrating firmer sediment suffer least, and open burrows may even remain unfilled (Fig. 8.11). The difference in the nature of burrows fills (Fig. 8.1) renders them susceptible as sites of preferred diagenetic mineralization. This can range from simple cementation of the fill, because of greater permeability, to pyrite formation, or in extensive silicification, as around the slender, deep burrow (*Bathichnus*). Loss of morphological detail due to compaction and often early and late diagenetic processes must be distinguished from the grain size or the manner in which the rocks split or break – the toponomic effects. When morphological detail is lost through such effects, identification may be uncertain in the same way that shells may be poorly preserved. With trace fossils any loss may render a more complex trace to have the features of a simpler burrow, part of a taphonomic series, or *taphoseries* (Fig. 8.1(I)).

Since skeletal materials are not actually present in trace fossils, except as burrow linings, what happens after the formation of a trace fossil depends on the state of the sediment at the time, its compaction and diagenetic history, together with the way in which any burrows have become filled. Or, if epigenic, what sedimentation and/or erosion modified the trace or print. Also important is the nature of the burrow margin. In most cases the margin was impregnated by an organic glue, which promoted the formation of diagenetic minerals. This is clearest in the fragments of the U-form burrow of the bivalve *Solemyatuba*, Fig. 8.1 (B), where the margin is deeply impregnated. Burrows that have become diagenetically enhanced are called *elite* traces.

A particular problem comes with heterolithic sediments, such as layered muds and sands or bioclastics, ooid-rich rock and micrite. Animals passing from one sediment into another will tend to modify their behaviour, if they are able to do so. Certain callianassid crustaceans need not line their burrow if the sand is sufficiently firm. Or only the roof may require lining. This is a bioengineering problem. But animal behaviour is not necessarily logical. Thus, in Tertiary sands near London with successive colonizations of *Ophiomorpha*, one might expect lined burrows to be crossed by unlined burrows as sediment has aggraded and become firmer. However, observations indicate that lined galleries may pass into unlined galleries, and lined galleries are crossed by lined galleries. Curiously the shafts are often unlined.

Burrow diagenesis is of particular concern to the permeability and porosity of the sediment (Section 8.5.3).

Box 8.1 Upper Jurassic rocks of the Dorset coast

The Dorset coast of southern England is likely to be appointed a World Heritage Site for its beauty and natural history. Geologically, it displays an almost unbroken section from the late Triassic–Rhaetic marine transgression, through the Lower Jurassic, which yielded the magnificent marine reptiles to Mary Anning, up through Middle Jurassic muds, sands and limestones to long reaches of Kimmeridge Clay, the main source rock for North Sea oil, and still further into the much-quarried Portlandian limestones of the Isle of Portland.

The Osmington Mills coastal section (Figs 8.5 and 8.6) in the Upper Jurassic is one of the most visited sections because of the diversity of sediment types seen in a short length of coast. Storms keep the section relatively clean and the section has received more than its fair share of research, not without controversy.

However, there is a big disadvantage with the Dorset coast section: only a single profile is available, but environments are three dimensional! A 5.5 m thick unit of sand has received particular attention, the Bencliff Grit. The interpretations that have been advanced include intertidal, estuarine, fluvial, washover (lagoon) and shoreface. They can't all be right! Another study has recognized three parasequences, with sequence boundaries at the base and top of the Bencliff Grit. The only way forward is to adopt an integrative approach, including an examination of the trace fossils and the palynology.

An MSc class in sedimentology attempted to wrestle with the problem. Equipped with hard hats, because of falling rock, they decided to map a section of the cliff. Working in pairs, each pair mapped two metres at a scale of 1 : 20 using graph paper, checking that their section tied in with that of their neighbours. This took a good three hours.

In the discussion that ensued it was clear there were four facies, and these facies displayed an ordering different from that given by many previous workers, What was paying off was the attention given to the muddy intervals, even though many were of short lateral extent due to penecontemporaneous erosion. The first point to emerge was that there were two 'types' of sand 'packages' and that both represented

Fig. 8.5 (a) View of cliff section in Bencliff Grit (Upper Jurassic), looking east to show dominance of fine-grained, oil-stained sands (facies A). (b) Detail of section photographed approximately 28 m to east of sketch (see Fig. 8.6), showing recovery succession above facies (A), passing successively to blocky silt and more muddy silt (B), to flaser-laminated sand and mud (C) and bioturbated muddy sand (D), which in turn is erosionally truncated by succeeding unit of facies (A). Scale bar graduated in centimetres.

Box 8.1 (cont'd)

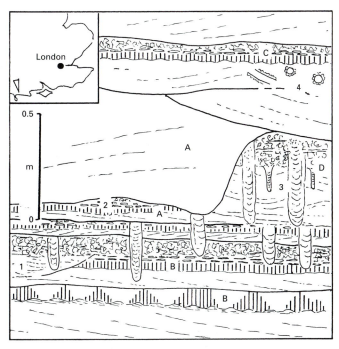

Fig. 8.6 Section of Bencliff Grit at Osmington Mills, Dorset, to illustrate relationships between the four facies: (A) (open ornament) cross-stratified fine-grained sand; (B) (vertical shading) unstratified mudstone/siltstone; (C) (lenticular ornament) flaser-bedded mud and silt/sandstone; (D) (scribble ornament) heterolithic and bioturbated sands and muds together with storm-event beds. Erosion at facies contacts and *Diplocraterion parallelum* and *D. habichi* are shown diagrammatically. (1) Amalgamation of facies A with storm-event bed in facies D; base of storm bed locally erosive, channelling into other facies. (2) Veneers of facies A followed by facies B and C, then truncation to facies A. (3) Thick event bed of facies D with top colonized by *Diplocraterion habichi*; *D. parallelum* penetrates to all facies but always originates in facies D. (4) *Ophiomorpha nodosa* in facies A, but has origin from a subsequently eroded higher level of facies D. Adapted from Goldring *et al.* (1998).

'event' sedimentation. When this was appreciated, it became clear that the bioturbated sands and muds with *thin storm sand beds* and a diverse trace fossil association (facies D of Fig. 8.6) represent normal (or close to normal) marine shelf sedimentation, into which there were major incursions of fine sand (facies A, which accounts for 85% of the unit), devoid of an autochthonous biota. These sands were deposited by a high-density current as can be seen from facies B (which always follows); several centimetres of muddy silt with finely divided plant debris (coffee grounds) set at all angles to the stratification, and representing deposition from the decelerating flow. Each event of facies A sedimentation must have eliminated the biota over an appreciable area. There is a gradual change to the 'recovery stage', facies C (flaser-bedded grey mud and sand with occasional trace fossils) and then to facies D.

Several depositional models were suggested. To help constrain them, a number of samples were taken for micropalaeontological analysis (foraminifera and palynofacies). The location of each was marked on the sections drawn, which would be useful for the micropalaeontologists and for subsequent interdisciplinary discussion. Although efforts were made to collect fresh samples, by digging into the clayey bands, the reports received later showed that some of the samples were too weathered and that some of the sampling would have to be repeated. The trace fossils were described some years ago, though unfortunately, before the sedimentological work. Now it was possible to link the ichnotaxa to likely colonization surfaces, reversing several earlier interpretations.

Perhaps the most useful conclusion to be drawn was this: the only way forward is to adopt an approach as fully integrated as possible, including the sedimentology, palaeontology, ichnology and micropalaeontology, and to sample with close relation to the observations of the sedimentologist and ichnologist.

Source: Goldring *et al.* (1998)

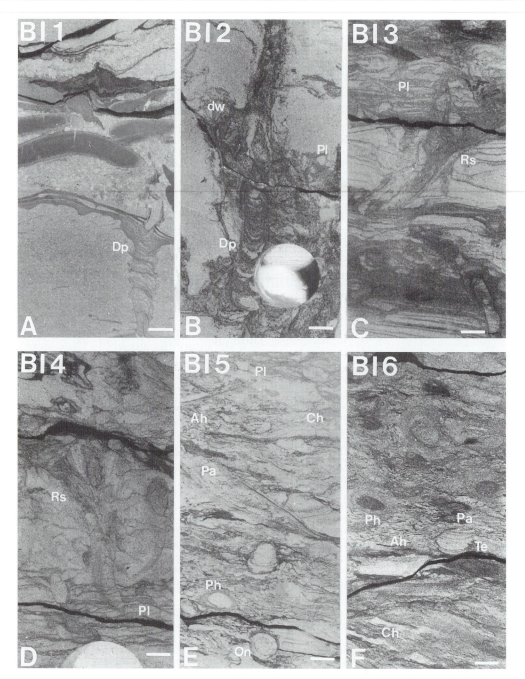

Fig. 8.7 Degrees of bioturbation and Bioturbation Index (BI) (Table 8.2(B)). (A) *Diplocraterion* ichnofabric with sparse bioturbation (BI = 1) with a discrete burrow of *Diplocraterion parallelum* (Dp) that has been truncated. (B) *Diplocraterion* ichnofabric with low bioturbation (BI = 2) with a large burrow of *D. parallelum* (Dp), which has been reburrowed by small *Planolites* (Pl). Dewatering (dw) has taken place around the burrow margins. (C) *Rosselia* ichnofabric with moderate bioturbation (BI = 3) with two colonization events just overlapping, both with *Rosselia* (Rs) and *Planolites* (Pl). (D) *Rosselia* ichnofabric with high bioturbation (BI = 4) where a number of colonization events by *Rosselia* (Rs) and *Planolites* (Pl) have overlapped. (E) *Phycosiphon* ichnofabric with intense bioturbation (BI = 5), which consists of a diverse ichnofauna of *Phycosiphon* (An), *Planolites* (Pl), *Palaeophycus heberti* (Pa), *Phoebichnus* (Ph), *Chondrites* (Ch), *Ophiomorpha nodosa* (On) where reburrowing and burrow overlap is common. (F) *Phycosiphon* ichnofabric with complete bioturbation (BI = 6), where no primary sedimentary structures are visible due to a complex history of burrowing. *Phycosiphon* (An) is cross-cut by *Palaeophycus heberti* (Pa), *Teichichnus* (Te) and *Phoebichnus* (Ph) which is reburrowed by *Chondrites* (Ch) and *Anconichnus*. Scale bars = 10 mm. Material is the vertical cut-section of borehole cores from the Lower Jurassic Tilje Formation (Båt Group) (A–E) and the Ror Formation (Fangst Group) (F) taken on the Mid Norwegian Continental Shelf.

8.3 Ichnofabric

So often, at outcrop or in core, the sedimentologist is faced with bioturbated sediment with a minimum of primary sedimentary structures preserved. What can be done? The primary fabric is well-nigh lost and a new (ichno)fabric imparted (Figs 8.7 and 8.8). This is the result of animal activity on all scales (and, of course, roots in soils). The situation can be so serious that one is tempted to turn away in dismay, but it is important to try and understand how it all came about. This can be done by making an integrated examination. Begin with the primary structures (if any remain), estimating the grade of bioturbation (bioturbation index) and then mapping the ichnology

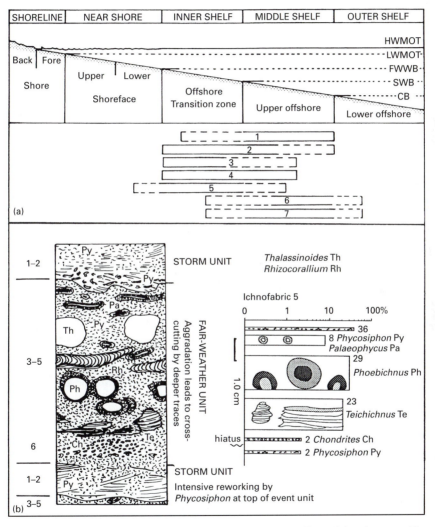

Fig. 8.8 (a) Inferred distribution of ichnofacies associated with *Phycosiphon incertum*. The seven ichnofabrics relate to sedimentation rate, substrate consistency, depth, storm wave base (SSB) and base of storm generated currents (CB); the depth is related to the fair-weather wave base (FWWB). (b) Schematic diagram of ichnofabrics 1–6 in relation to event bed and fair-weather (equilibrium) processes, with schematic ichnofabric constituent diagram of ichnofabric 5, BI = 6 (Fig. 8.7(F)), where *Phycosiphon* in muddy, fine-grained sandstone is partly obliterated by deeper burrows. The horizontal axis plots the percentage area of the sedimentary structures and burrow types on a log scale. The vertical axis plots events (in order) from the initial sedimentation events (primary fabrics) to subsequent modification by bioturbation (secondary fabrics). Adapted from Taylor and Goldring (1993).

Table 8.2 Bioturbation and sedimentary facies.

A OBSERVATIONS AND IMPLICATIONS

Observations	Implications
Lamination destroyed	An averaging of overall grain size
Grain orientation destroyed	Overall reduction of porosity and reduced permeability
Coarse and fine laminae mixed	Sediment maintains higher water content
	Lithification delayed
Fines selectively removed and passed into suspension	Increase in grain size, porosity and permeability
Fines moulded into faecal pellets	Sediment grain size effectively increased
Sediment trapped by tubes or plant stems	Decrease in grain size
	Stabilization of substrate
Sediment bound by roots	Stabilization of sediment
Vertical burrows	Burrows act as drainage channels, hence sediment firmer than expected
	Permeability normal to bedding increased
Bedding-parallel burrows	Permeability normal to bedding probably reduced
	Permeability parallel to bedding may be enhanced
Burrow margin lined	Preferred permeability pathways
Open burrows vastly extending sediment–water interface	Substantially increased geochemical fluxes below substrate

B BIOTURBATION INDEX[a]

Bioturbation index (BI)	Fraction bioturbated (%)[b]	Classification
0	0	No bioturbation
1	1–5	Sparse bioturbation: few discrete traces and/or escape structures
2	6–30	Low bioturbation: bedding distinct, low trace density, escape structures often common
3	31–60	Moderate bioturbation: bedding boundaries sharp, traces discrete
4	61–90	High bioturbation: bedding boundaries indistinct, high trace density with overlap common
5	91–99	Intense bioturbation: bedding completely disturbed (just visible), limited reworking, later burrows discrete
6	100	Complete bioturbation: sediment reworking due to repeated overprinting

C INTERPRETING SEDIMENTARY FACIES

(1) Identification of major environments

- Non-marine fluvial and floodplain facies
- Marine and marine-influenced (marginal) facies
- Deep-water muds (marine and non-marine)
- Event beds (storm, tempestite, turbidite, etc.)

(2) Key stratal surfaces

- Firmgrounds, hardgrounds, omission surfaces

(3) Genetically related succession

- Depth related in neritic facies
- Depth (illumination) related in hardgrounds

- Proximality related in turbidities, storm and tempestite event beds
- Oxygenation trends
- Salinity fluctuations in lagoonal facies

Correlation with Seilacherian ichnofacies

- Fluvial and floodplain: *Scoyenia*
- Neritic: *Skolithos, Cruziana*
- Deep-water: *Nereites* (in part), *Zoophycos, Mermia*
- Event beds: *Arenicolites, Nereites* (in part)
- Firmgrounds: *Glossifungites*
- Hardgrounds: *Trypanites* (*Teredolites*), *Entobia, Gnathichnus*

[a] Each grade is described in terms of the sharpness of the primary sedimentary fabric, burrow abundance and amount of burrow overlap. Adapted from Taylor and Goldring (1993).
[b] Use these percentages as a guide, not as an absolute class division.

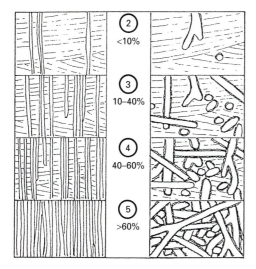

Fig. 8.9 Diagrams to assess degree of bioturbation for *Skolithos* (left) and *Ophiomorpha* (right) Each frame represents an area 50 × 35 cm normal to bedding. Separate diagrams must be constructed for other types of ichnofabric. Grade 1 = no bioturbation; note that no indication of tiering is expressed. Adapted from Droser and Bottjer (1987, 1989).

(Section 8.10). Since ichnofabric integrates the diagenetic, stratinomic and ethological aspects with the sedimentology, it is likely to be significant. Trace fossils in cores present special difficulties, in that it may not be easy to see the nature of the bioturbation on the rough outside or sawn surfaces, or because of the film of drilling mud, which may penetrate the core. Cores may be oblique to bedding. Actual bedding information may be difficult to discern. But there is often the advantage that muddy and sandy sediment core equally well.

To simply state that a sediment is bioturbated is like saying that a sediment is fossiliferous – it does not convey much. Categories such as low and high have little meaning. Two schemes for the assessment of bioturbation are in current use: Taylor's bioturbation index (Table 8.2) aims to relate the degree of bioturbation with the preservation of the primary sedimentary structures. This needs a little care but works well. The area (or length of core) classified must be noted. Taylor's scale is based on the Reineck scale developed with long experience of modern sediments. Droser and Bottjer's scheme (Fig. 8.9) is rather simpler but only considers the degree of bioturbation, as the ichnofabric index. There is no zero – ii1 (ichnofabric index 1) indicates no bioturbation – and the scaling is different.

The *ichnofabric constituent diagram* (Fig. 8.8) provides a visual means for describing and comparing ichnofabrics by (1) assessing the extent and type of the remaining primary fabric, (2) unravelling the tiering and ordering of the ichnotaxa (Section 8.5.6), and (3) recognizing the type of bioturbational event. It can also be applied to firmgrounds and hardgrounds. For convenience, burrow diameter is used. The test of a satisfactory construction is to ask someone to make a drawing of the rock from your diagram!

8.4   What do trace fossils mean?

Animals construct burrows in four ways. How they do this is discussed in detail by Bromley (1996). An initial probe may be extended in several ways:

- Enlargement
- Pushing ahead
- Moving sideways
- Probing laterally

These methods may be used in combination and in various planes relative to the substrate surface. Refinements (that can be modelled on software) include random progression, meandering, coiling, together with margin lining and backfilling. Some of the complexities are shown in Fig. 8.10, and the behavioural purpose they serve.

8.4.1   Trace fossils and behaviour (ethology)

Nine basic patterns of animal behaviour (Figs 8.10, 8.12) are generally recognized, patterns which describe the morphology actually registered in a trace fossil. The animal may also have been involved in other activities, e.g. feeding while resting. But in most cases we cannot detect any evidence of feeding or what trophic (feeding style) was taking place (Box 8.2). In contrast the drill-hole in a bivalve shell (Fig. 5.8) is clear evidence of predation (praedichnion) but the predator gastropod or octopus has not left any other evidence of its identity, though fortunately carnivorous gastropods are still very much around! There is also an overlap and integration between most categories; this is especially marked with burrows indicating wandering and sediment feeding. A further complicating factor is that we do not understand the purpose of many traces, e.g. *Zoophycos*.

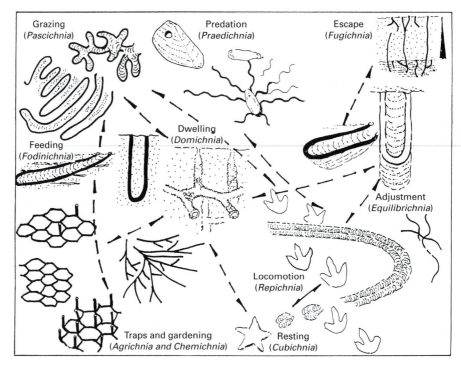

Fig. 8.10 An ethological classification of trace fossils with nine styles of behaviour. Arrows suggest some interrelationships. (Clockwise from top right) Escape structures made by polychaete *Nereis* through an (event) sand bed. *Diplocraterion* showing adjustment downwards (protrusive) and then upwards (retrusive). Locomotion traces (arthropod, trilobite), bipedal vertebrate track, irregular burrow (*Planolites*). Resting traces (*Rusophycus*) and starfish. Complex feeding traces of *Chondrites* (?chemosymbiont), *Paleodictyon* with two behaviour modes: (below) with multiple outlets to serve for 'gardening' and microbe culture; (above) with few openings to serve as a trap for motile micro-organisms. Sediment feeding in *Rhizocorallium*. Dwelling burrows *Arenicolites* and *Ophiomorpha*. Systematic foraging, deposit feeders. Meandering *Helminthoida* and antler-like *Phycosiphon*. *Crassatella* with predatory naticid gastropod drill-hole (see Fig. 5.8), and lower surface of a thin sand bed with positive hypichnia of annelid traces (*Cochlichnus*) converging on a shallow burrowing bivalve. Scales variable.

Seilacher, in introducing the scheme, gave a Latinized name to each category. The most important relate to dwelling, feeding and movement:

- Resting traces (Cubichnia): temporary, stationary position.
- Dwelling traces (Domichnia): permanent or semi-permanent abode.
- Locomotion traces (Repichnia): includes walking, crawling, tunnelling, running, swimming (Natichnia).
- Adjustment traces (Equilibrichnia): slight upward or downward movement by animal to maintain equilibrium with the substrate surface.
- Feeding traces (Fodinichnia): deposit feeding combined with dwelling.
- Escape traces (Fugichnia): escape from burial by sediment influx, or toxicity. The sedimentary evidence for escape should be clear.

- Grazing traces (Pascichnia): systematic (programmed) exploitation of the substrate surface or substrate for food, involving avoidance of previous burrows (phobotaxis); compare this with field ploughing.
- Intense farming (gardening), traps (Agrichnia): not an altogether satisfactory grouping, which includes microbial traps, microbial culturing, and ranges into microbial chemosymbiosis (Chemichnia).
- Predation traces (Praedichnia): drill-hole (*Oichnus*) made in shells by carnivorous gastropods and octopus, crab predation.

Other categories have been proposed for subaerial structures, such as termite nests and insect breeding structures (Calichnia) in dune sands. In the field it is useful to plot a pie diagram to show the proportion of different ethological groups present in different facies.

Box 8.2 Lower Cambrian trace fossils

How does a graduate choose a project for postgraduate work? To have the opportunity to work on one of the world's most important fossiliferous sites is an almost unimaginable situation for most graduates. Sören Jensen had hoped to be able to work on fossil fish and actually had a dislike for trace fossils. But Stefan Bengtson, one of Sören's teachers at Uppsala University and later his supervisor, suggested a small topic for his basic palaeontology course: trilobite burrows that followed worm burrows. Were the trilobites hunting? His interest was aroused and, on graduating, the opportunity came to make a study of the ichnology of the 10 m thick Lower Cambrian Mickwitzia Sandstone of central Sweden. A good many papers had already been written over the past century and large collections were available in national museums and with amateur collectors.

The sediments are unmetamorphosed and rest horizontally on the near-planar unconformity with Precambrian gneiss. Exposures are limited and the best are in small mines where the Cambrian sediment was dug to gain access to the weathered gneiss; the gneiss was extracted for millstones and exported throughout Europe. As might be expected, the Cambrian sandstones do not contain much in the way of body fossils, and early Cambrian trilobites appear first in the overlying beds. The Mickwitzia is of Lower Cambrian age, but not lowest.

Judging from the number of different trace fossils (including those due to trilobite activity), there must have been a sizeable biota present. Sören wondered how this could be measured, 'Didn't many of the traces, for which different names had been used, just reflect different behaviour and/or different preservations?' As may be seen in Chapters 5 and 8, these are questions that palaeontologists and sedimentologists will have to consider more carefully. At present, it often seems to be a case of the more names I can use, the more I can impress my colleagues!

Sören also has had to contend with a diverse collection of structures of doubtful organic origin, often known as dubiofossils. In his monograph, Jensen illustrated a number of these dubiofossils and some have been judged to be of inorganic origin. The only way to determine this is to observe specimens that are actually in the rock and analyse their stratinomy. And then, if necessary, to slab (this is of course destructive).

Source: Jensen (1997)

Fig. 8.11 (Left) *Cruziana rusoformis* (Lower Cambrian, Sweden): plan view (upper) and lateral view (lower) to show truncation event and overlying rippled surface. (Right) Limestone below a hardground where burrows have remained open (Middle Jurassic, Gloucestershire, England). Scale bars = 10 mm.

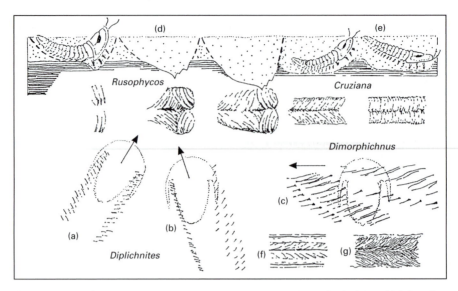

Fig. 8.12 Trace fossils made by trilobites to illustrate a variety of principles. Multifunction, differences in trace produced by an animal: (a) walking straight ahead, (b) slightly obliquely, (c) sideways, (d) making a short burrow, (e) continuous plough. Polynomial nomenclature reflecting different behaviours; preservational variation: (d) it depends on the depth to which the animal excavated in mud below the sand layer. Variation due to different attitude: (e) during burrowing. Stratigraphic value: (f) *Cruziana semiplicata* (Upper Cambrian), (g) *C. furcifera* (Lower Ordovician). Ichnogenera are separated by distinct differences in behaviour. Preservational differences are not normally of taxonomic significance. Adapted from Seilacher (1970).

Relationship between ethology and trophic structure

Studies on modern environments show that there is no simple relationship between feeding mode and trace morphology. There are two major uncertainties for palaeoichnology: ethological interpretation of the trace (how and why it was constructed), and the different modes of feeding that can be used for identical structures. For instance, it could be a fossil U-burrow (*Diplocraterion* or *Arenicolites*), representing a permanent or semipermanent dwelling (domichnion), which might serve for suspension feeding, a site for predation or detrital feeding. (It did *not* serve for sediment ingestion as does the J-form of the lugworm burrow *Arenicola*.)

Nevertheless, some generalizations are useful, with care. Muddy substrates tend to be associated with deposit feeding, and the transition to a firmground or hardground signifies a change to suspension feeding. Sandy substrates are favoured by many suspension feeders. Mixed heterolithic sediments tend to have the most diverse suite of trace fossils, reflecting infaunal and epifaunal suspension and deposit feeding. The problem of interpreting the complex and often deep structures of *Chondrites* remains, Does it always represent chemosymbiosis? Referring to *Zoophycos* (Figs 8.4 and 8.13), Bromley (1991) posed the question, Does it represent strip-mining by a deposit feeder, a backfill derived from the surface (waste stowing), the most favoured interpretation, a food cache for future use (nutrient supply in the deep sea is highly periodic), or a food cache to be subsequently 'farmed'?

Looking at bioturbation over geological time provides information about the evolution of behaviour and also a limited amount of biostratigraphic data (Section 8.6). Another complicating aspect is that, in the Palaeozoic, *Zoophycos* may represent the activity of an opportunist, whereas in the Mesozoic, K-strategy is represented and with a more complex behaviour pattern.

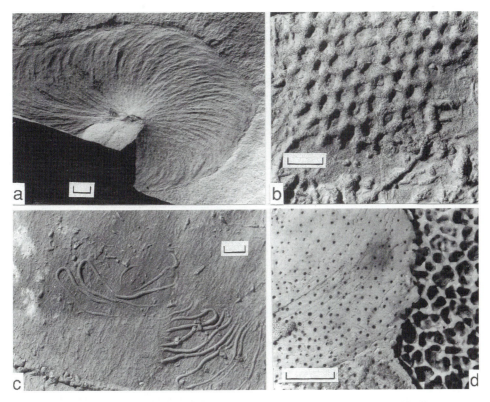

Fig. 8.13 (a) *Zoophycos brianteus* in bedding-parallel section (Helminthoid Flysch, Middle Eocene, near Palma, Italy); (b) *Paleodictyon minimum* as predepositional convex (positive) hypichnion on sole of turbidite (eastern Turkey); (c) turbidite sole with partly eroded preturbidite (positive hypichnion) *Helminthoida crassa*, and post-depositional *Strobilorhaphe clavata* (Zollhaus, near Fribourg, Switzerland, Palaeocene); (d) scallop shell (*Pecten*) partly split to reveal cavity system (*Entobia*) caused by sponge *Cliona* (Recent, Newfoundland). Scale bars = 10 mm.

8.5 Colonization and ecological factors

Every trace fossil at the time of its construction related to a substrate surface. Where was this surface in the rock section and what ecological factors influenced its colonization? These are perhaps the most important questions faced in the field, but they are often difficult to answer. Identification of the colonization surfaces is particularly important for

- Deep burrowers
- Event beds
- Omission surfaces (Fig. 8.28, p. 161)

Event beds include turbidites and storm event beds in marine and non-marine facies. Consider a deeply

burrowing, callianassid shrimp pellet-lining its burrow to form the trace *Ophiomorpha*; it can penetrate down to 4 m below the surface and into a different facies. Likewise *Chondrites* may extend down to about 1 m. In the first example, water in the burrow and above was near to normal salinity but the sand adjacent to the burrow could have been deposited in a river. It was just sand! Furthermore, there is a high probability that the actual surface colonized was aggraded and then subjected to erosion, all during burrow occupancy. This illustrates the problem of using the traces of deep burrowers for environmental interpretation without ascertaining where the substrate surface was. A problem is posed in recording such occurrences on graphic logs and which are then to be used in facies interpretation.

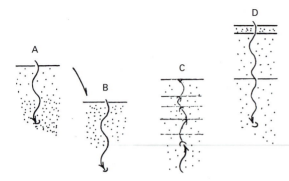

Fig. 8.14 Four colonization situations: (A) from the normal top of a bed; (B) from an eroded surface (firmground); (C) from an underlying level by keeping pace with aggradation; (D) from an appreciably higher level.

With the traces of shallow burrowers, it is easier to appreciate where the substrate surface was, and the ecological parameters involved can be more easily assessed. This means working out the type of bedform that was colonized.

8.5.1   Colonization strategy

Any burrow we see may have been initiated at one of four possible sites (Fig. 8.14):

1 Can the substrate surface colonized be positively identified, e.g. funnel opening, complete bivalve crypt?
2 The surface may have been penecontemporaneously eroded and/or lithified before colonization. Can the amount of erosion and lithification be estimated (Fig. 8.21, p. 152)?
3 The trace may be due to an organism entering the sediment by moving upwards. Was the animal responsible, escaping or adjusting its position in response to gradual accretion (Section 8.5.4)?
4 The trace may be the result of an animal burrowing down deeply from a higher level (as discussed above).

There are four strategies of colonization. *Opportunistic*, rapid colonization of a vacant ecological niche: as following an overbank spill on the floodplain, or deposition of a storm event bed, or tephra (volcanic ash) layer (Fig. 8.30, p. 163). Opportunists are of low diversity and tend to be of small size, and hence with generally small diameter burrows. Their life span also tends to be short, so do not expect large, complex burrow systems. Providing another event does not return conditions to square one, the 'generalist'

pioneers will be followed by a more diverse biota leading to an *equilibrium* situation and then to a *climax*, mature community. The mature community may include complex burrow systems, e.g. *Ophiomorpha* boxworks, *Zoophycos*.

Colonization may be initiated by *larval spatfall* or by *relocation*, or by a combination of the two. The colonization processes in deeper water are virtually unknown but it is generally considered that large burrows found in association with turbidites were formed by animals introduced by the turbidity current from shallower water. They tolerated the new situation but did not survive 'normal' sedimentation and probably never reproduced. They are most common in proximal turbidites.

An *absence of bioturbation* is one of the problems most frequently encountered at outcrop or in core. Is the sediment truly devoid of any primary sedimentary structures or bioturbation? If the sands are cross-stratified or the muds well laminated, it is reasonable to conclude there was no opportunity for colonization. If there has been bioturbation then there is nearly always some evidence – slight mottling or mud streaks that could not be due to hydraulic processes. If such details are absent and the use of various techniques (Appendix A) fails to help, then it is best to conclude that the sediment was not bioturbated. One can then consider the physical and chemical factors that prevented colonization.

In many environments the substrate was colonized only intermittently and then only to create a scattering of traces. Excluding plants, this is the case in fluvial environments and floodplains, and deltaic environments.

8.5.2   Substrate classification

Trace fossils and bioturbation provide the clearest indications for the state of the substrate (soft, firm, hard, etc.) at the time of their construction. And also for any sequential changes to the substrate that might have taken place. These are particularly important in the recognition of omission and key stratal surfaces (Fig. 8.28, p. 161). A general indication of the response by organisms to substrate conditions is shown in Fig. 8.15 and the five main states and their identification in Table 8.3. In modern mud-rich sediments the nature of the bioturbated substrate is covered by the terms *mixed, transition* and *historical* layer (Fig. 8.16). *Mixed* layer refers to total bioturbation and uniform colour and may be very thin in sandy sediment. The *transition* layer shows a tiered struc-

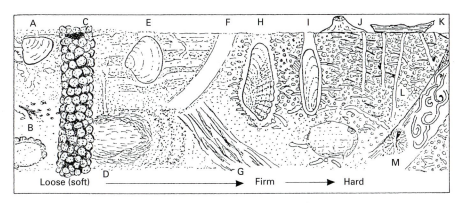

Fig. 8.15 Diagram to show types of biogenic response to loose (soft), firm and hard (lithified) substrates (approx. ×0.5 unless otherwise indicated): (A) *Nucula*; (B) *Chondrites*; (C) *Ophiomorpha*; (D) *Echinocardium*; (E) *Mercenaria/Arctica* (×0.25); (F) *Thalassinoides*; (G) *Spongeliomorpha* (with scratched burrow margin); (H) *Petricola*; (I) *Lithophaga* (insoluble grain on left side) (×1.0); (J) crinoid hold-fast; (K) oyster: (L) *Trypanites*; (M) encrusting bryozoan and serpulids on cemented burrow margin.

Table 8.3 Substrate type, observations and identification.

Observations	Deductions	Type of ground
In mudrocks and micrites, sediment highly diffuse, possible traces highly compressed and smeared, feeding burrows dominant, e.g. *Planolites*	Sediment was more or less water-saturated, compaction extreme	Soupground
In muddy sediment burrows show substantial compaction and indefinite margins Bioglyphs indistinct or not preserved Mixed feeding (farming traces in deep water), micrites, e.g. *Chondrites, Zoophycos, Phycosiphon, Helminthopsis*	Some dewatering	Softground
In sandy sediment, burrows lined Bivalve burrows, *Arenicolites*	Permanent burrows, requiring stabilized margins	Looseground
Burrow outlines sharp, and may have distinct sharp bioglyphs (scratches) ± slight deformation of crypts and compaction of burrows: *Spongeliomorpha, Glyphichnus* Some encrusters, e.g. *Liostrea*	Stiff, dewatered substrate	Firmground[a] Concealed firmground[b]
Mineralized crust (±) Boring and encrusting biota Borers cut evenly across matrix and skelelal grains	Lithified substrate surface	Hardground (in limestones) Shellground (cemented shell bed) Rockground (with tectonic omission)

[a] Firmground is not a well-defined category and is wide-ranging.
[b] Sharp change in behaviour, generally associated with lithological change, indicating firm interface *within* the *transition* zone. The effect tends to produce burrowing along the interface. Look out for differences on upper and lower sides of traces, and indications of non-penetration.

ture (Fig. 8.24, p. 155) of distinct burrows and marked colour contrasts, associated with variable oxygenation of the sediment; in fact, this is a diagenetic effect determined by the bioturbation. The *historical* layer reverts to a more uniform colour and it is this layer that will normally become fossilized. Most formerly open burrows in muddy sediment will have become closed and pellets compressed. The consistency of

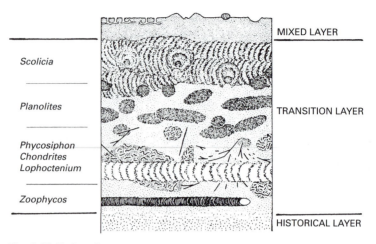

Scolicia

Planolites

Phycosiphon
Chondrites
Lophoctenium

Zoophycos

MIXED LAYER

TRANSITION LAYER

HISTORICAL LAYER

Fig. 8.16 Tiering diagram constructed from deep-sea box cores in muddy sediment to show distribution of *mixed* (homogenized) and *transition* (excavation, backfills and spreite) layers, to *historical* (fossilized) layer. Surface trails and the open burrows of *Paleodictyon* are shown in the mixed layer. Adapted from Wetzel (1984).

the sediment, even within the transition layer, shows changes in physical properties to which burrowing organisms may or may not be able to respond. This is where *concealed firmgrounds* are particularly significant.

Firmgrounds are not readily identified and it may be possible to recognize degrees of firmness. Hardgrounds with borings and encrusters are readily recognized (Box 8.3). Gruszczynski (1986) showed that very few hardgrounds in shallow-marine settings had undergone gradual lithification, the evidence for which would be a succession from burrowers to borers and encrusters: *gradual ecological replacement.* Many show successive planation events due to wave action with coarser particles systematically grinding down the surface. In shallow-marine environments small shifts in relative sea level can lead to emergence, frequently accompanied by incursions of meteoric water and rapid cementation. Wilson and Palmer (1992) discussed the formation of modern hardgrounds, cement minerals and changes over geological time.

Unconformity surfaces and exhumed surfaces are rockgrounds: the biota colonized a surface following a (major) stratigraphical hiatus. (But the surface may not have become lithified!) Borings may penetrate and encrusters overlap cemented faults and fissures. Skeletal substrates also display such overlap. Hiatus concretions can be encrusted *in situ* and/ or subsequently, when broken and incorporated into a conglomerate.

Although hardgrounds formed over a wide range of environments there seem to be two main settings: (1) in shallow, warm epeiric seas, and (2) in somewhat deeper offshore water, as in Cretaceous chalk of northwest Europe. In the first the hardground is typically abraded and planar, with lateral extent ranging from a few metres to several hundred kilometres. In the second it is irregular, due largely to the bioturbational topography produced in an environment of relatively low energy; a sediment-starved (abandoned) surface. The distribution of the biota is contrasting: in the first, there may have been several colonizations with intermediate periods of planation. In the second, upper surfaces are sparsely encrusted and bored because of dustings of fine-grained sediment.

Hard substrate borings show a depth zonation because they were made in the following ways:

- Shallow-water bivalves such as *Lithophaga* (constructing *Gastrochaenolites*, Fig. 5.19).
- Species of boring sponge that have symbiont algae (some *Cliona* constructing *Entobia*, Fig. 8.13).
- Species of *Cliona* that are depth-restricted.
- Echinoids that feed on algae and mark the rock (*Gnathichnus*).

Microborings, reviewed by Perry (1998), also show depth zonation. The distribution of borings provides the only good evidence for actual depth. Geologically this can be applied to lateral and vertical changes such as occur in tectonically active areas, e.g. the Mediterranean (Bromley, 1994).

Box 8.3 An Ordovician hardground

Hardgrounds (Fig. 5.18) are of particular interest to palaeontologists because they have a special story to tell of a colonized, cemented rock surface. In some cases a faunal continuum can be demonstrated, reflecting the gradual change in cohesiveness of the seafloor and its planation by sand under wave action. Tim and Caroline Palmer gave the first detailed account of an Ordovician hardground and its community structure.

At least 100 m² of hardground was exposed in a road-cut in northeast Iowa, but the surface had been locally damaged by road machinery. An area some 3 m² was found to be undamaged and was marked with a water-soluble marker into 30 cm² areas (Fig. 8.17 and Appendix B), and the encrusting fauna accurately plotted on a half-scale plan. This took three days to accomplish, recording every element greater than 2 mm across. The encrusting and boring biota was composed of 2 types of borings, 3 types of crinoid holdfast and 11 types of encrusting bryozoan. One of the borings was new and it later required proper description, naming and assignment of type material to the National Museum of Natural History, Washington DC. In addition a diverse, non-cemented epifauna was collected from the shale immediately above the hardground, which included a sponge, coral, brachiopods, gastropods, other bryozoans, trilobites and a tube worm. Field identification of the bryozoans raised problems because some had not been described, and because of external homeomorphy and the small size and immaturity (rather than stunting) of the colonies. Arbitrary 'field' taxa were designated on the basis of growth form and surface characterization. Where possible, these were subsequently identified with the aid of acetate peels and thin sections.

The degree of detail was needed because the authors recognized the opportunity to apply work on modern hardground substrate faunas to the Ordovician, and to determine whether controls on larval settlement and intra- and interspecific competition could be recognized, and if so, whether they were similar to those acting today. This aspect was timely because much work was then being done on modern bryozoans and tube worms. Larval settlement on modern rocky surfaces is largely controlled by biotic factors.

Appendix B presents a different statistical method to that originally used by the Palmers. The uniformity of the Ordovician hardground suggests that the strong clumping observed was not due to any heterogeneity, though *Trypanites* (Fig. 8.18) did tend to be associated with low humps. Limited evidence indicated that mechanisms to avoid intraspecific competition were employed by Ordovician bryozoans, and destructive overgrowth between species was also observed. A difficulty in the analysis of interspecific competition in fossils is in establishing that the colonies lived contemporaneously. It was possible to draw up a diagram to show species dominance, which indicated that a high degree of interspecific competition already existed in the Ordovician.

Modern hardgrounds are generally fully covered, but the Ordovician hardground showed only an average 5% cover. This was attributed to the intensity of erosion destroying the biota, to which Ordovician taxa were much less resistant than those colonizing Mesozoic and younger hardgrounds.

Selected areas and colonies were photographed and specimens from outside the gridded area collected. More work can be carried out on the shale biota and on specific sedimentological aspects; for instance, on the diagenetic history of the hardground using isotope analysis.

Source: Palmer and Palmer (1977)

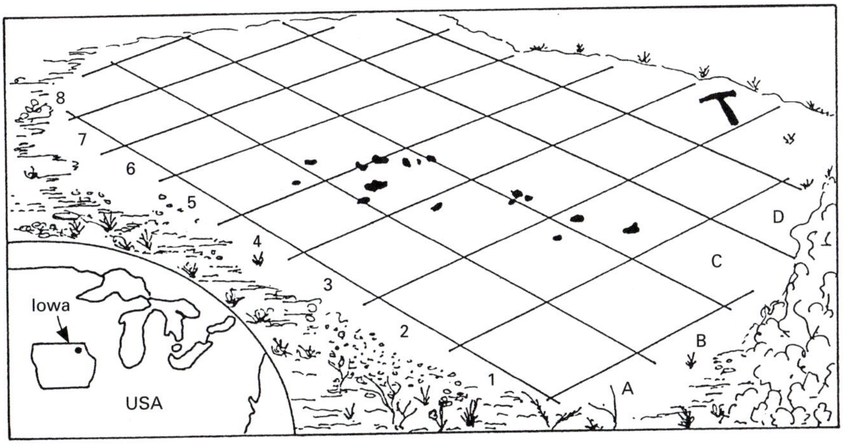

Fig. 8.17 Sketch from a photographic illustration of the Ordovician hardground in north-east Iowa analysed by Tim and Caroline Palmer. Blobs indicate the position of colonies depicted in Fig. B.1 (p. 175), where quadrat lines are omitted.

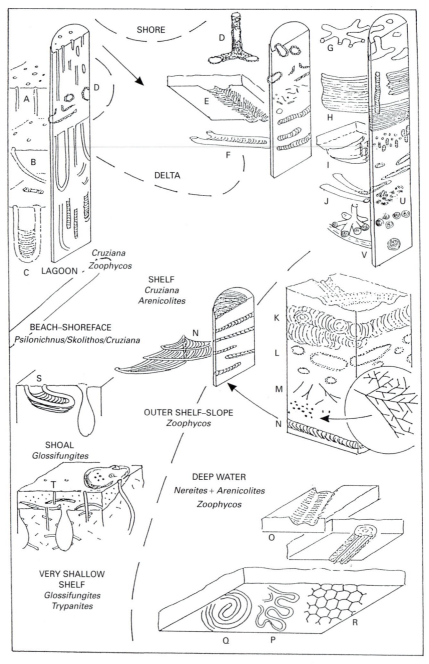

Fig. 8.18 Diagram to illustrate the distribution of the traditional (Seilacherian) ichnofacies, and sections of ichnotaxa as seen in split core. (A) *Skolithos*, (B) *Arenicolites*, (C) *Diplocraterion*, (D) *Ophiomorpha*, (E) *Cruziana*, (F) *Rhizocorallium*, (G) *Thalassinoides*, (H) *Teichichnus*, (I) *Phycodes*, (J) *Schaubcylindrichnus*, (K) *Subphyllochorda*, (L) *Planolites*, (M) *Chondrites*, (N) *Zoophycos*, (O) *Scolicia* (above), *Subphyllochorda* (below) exogenic and endogenic echinoid burrows, also shelf and intertidal, (P) *Cosmorhaphe*, (Q) *Spirorhaphe*, (R) *Paleodictyon*, (S) *Glossifungites* (= *Rhizocorallium jenense*), (T) *Trypanites* and *Gastrochaenolites*, (U) *Phycosiphon,* mantled, discontinuous faecal core, (V) '*Cylindrichnus*', concentrically laminated burrow, and *Asterosoma* (radial).

8.5.3 Grain size distribution, porosity and permeability

Grain size is generally readily estimated in the field. But a grain size distribution derived from bioturbated sediment can have no hydraulic significance. Likewise, in destroying primary sedimentary structures and the primary fabric, bioturbation also destroys the directional permeability associated with them. Strongly bioturbated intervals will normally have reduced permeability, although the mixing-in of a proportion of mud seems often to have hindered cementation. But if fines are expelled at the water column by deposit feeders (e.g. scaphopods), rather than being pushed aside, then the sediment will be cleaned. Sand-filled burrows penetrating heterolithic sediments can act as conduits between the sand layers. In general, vertical burrows may increase interstratal flow, and bedding-parallel burrows may increase stratal flow. Burrows formed at depth in firm sediment may have a loose fill, or even remain open (Fig. 8.11). This considerably enhances the permeability. In partly bioturbated sediments the primary grain size distribution can be measured if care is taken to sample only areas exhibiting primary sedimentary structures. (In lithified rock this can be done on a thin section; 50–100 grains may be sufficient.)

The manner in which animals penetrate the sediment is obviously much influenced by grain size and packing, as is the way animals feed on or manipulate grains. For example, some (edible) crustaceans are restricted to medium-grained sand because they have evolved the means to handle and clean grains of this size. Grain size is a major factor in larval settlement.

8.5.4 Colonization and sedimentation rate

The extent to which the primary fabric is destroyed by bioturbation depends on the sedimentation rate and size of the animals. Where sedimentation rate is low the primary fabric may have been completely destroyed, and replaced by a mottling. With care this can be described and interpreted.

Response to sediment aggradation and minor erosional events

In shelf and shallow-water sediments, minor events of sedimentation and erosion commonly occur, owing to fluctuating energy conditions (Figs 8.19 and 8.21). Organisms can often adjust to such events and,

Fig. 8.19 (Left) Storm event sandstones in neritic succession, displaying sharp soles and bioturbated tops: parallel → bioturbated (Northern France, Upper Jurassic); hammer shaft = 0.3 m. (Right) *Diplocraterion parallelum*; (left) protrusive form with complete U-burrow, (centre) upper part of protrusive form, (right) retrusive form. All three were truncated prior to deposition of capping sand (Upper Devonian, North Devon, England); scale bar = 10 mm.

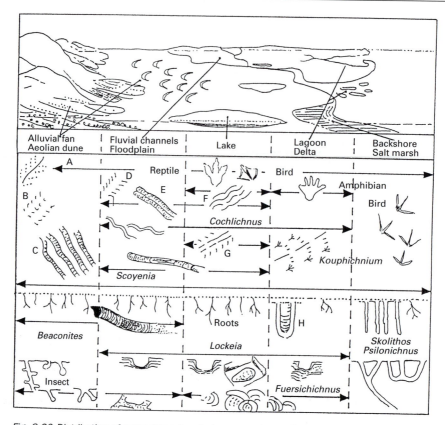

Fig. 8.20 Distribution of some trace fossils in non-marine environments with mainly epigenic (upper group) and mainly endogenic traces (lower group): (A) *Paleohelcura* (scorpion tracks), (B) *Mesichnium* (insect tracks); (C) *Entradichnus* (insect tracks and burrows), (D) *Acripes* (crustacean tracks), (E) *Isopodichnus* (phyllopod or notostracan burrow), also known by some as *Cruziana*, (F) *Undichnus* (fish traces), (G) *Siskemia* (arthropod track), (H) *Diplocraterion*, (I) *Psilonichnus* (crab burrow). *Beaconites* (centimetre-size are probably due to vertebrate; small due to insect or other arthropod). *Cochlichnus* (worm trail, burrow), *Kouphichnium* (xiphosurid tracks), *Scoyenia* (insect trail or backfilled burrow). Adapted from a personal communication of J. E. Pollard.

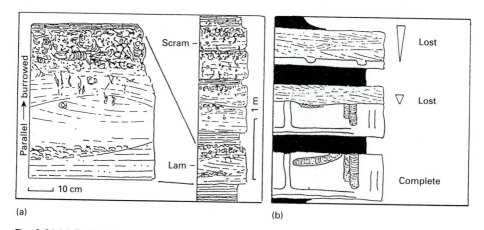

(a)   (b)

Fig. 8.21 (a) Typical bioturbation in shelf storm beds (either siliciclastic or calcareous): parallel → burrowed (lam–scram). (b) Trace fossils as indicators of amount of sediment lost by penecontemporaneous erosion. Adapted from Wetzel and Aigner (1985).

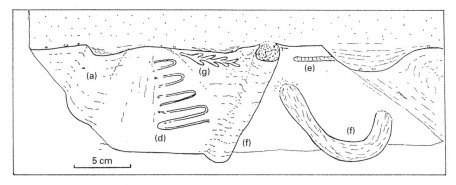

Fig. 8.22 Criteria to distinguish between pre- and post-depositional suites of trace fossils on turbidite soles (see text).

because so many trace fossils were formed at a high angle to the substrate surface in shallow water, it is often easy to detect vertical adjustment. Adjustment of a few mm–cm at any one stage can be accomplished by many bivalves, mobile echinoderms, anemones as well as annelids and arthropods. Only annelids, arthropods and gastropods can escape upwards through an appreciable sediment cover. In general, shallow burrowers respond to sedimentation better than deep burrowers. But the opposite is to be expected with deteriorating oxygenation.

Tidal flats adjacent to a vigorous channel may show little bioturbation because of the high frequency of small-scale aggradation and erosion. In quieter areas the likelihood of the sediment becoming bioturbated is much greater.

Response to major 'event' sedimentation

In major 'event' sedimentation, trace fossils provide convincing confirmation of an event that interrupted infaunal life. If it is clear that colonization took place only from the top of the bed and not from any level within, then there can have been only a single phase of sedimentation. Further confirmation in turbidites, if required, can be by analysis of the suite of sole traces associated with turbidites of different thickness. This also provides a measurement of the depth to which the culprits penetrated (Section 8.5.6). In thick event beds, only traces made by animals able to penetrate right to the sole, will be found. Take care to exclude amalgamated units (Fig. 6.2).

Trace fossils, although often distinct and diverse on the soles of *turbidites*, are usually few in number. Assess the assemblage present over a specified thickness. The best way is to record them on a log, noting whether the traces occur on the sole or top of each bed. The sole traces represent two suites (Fig. 8.22;

Kern, 1980): a *preturbidite suite*, those picked out by the turbidity current from the background mud and then infilled, and a *post-turbidite suite*, those which colonized from the top of the bed. In most siliciclastic turbidites it is rare to see burrows traversing a bed because of the nature of the sediment. If no erosion took place, the *preturbidite suite* may have been 'frozen'; this is particularly important for investigating the tiering.

Here are some criteria for distinguishing the *pre-event* suite:

- Erosional modification of traces by the turbidity current (Fig. 8.22(a)).
- Displacement of faecal strings.
- Traces on soles of very thick beds are likely to be predepositional, but beware of amalgamated beds.
- Partial erosion of laterally extensive traces (Fig. 8.22(d)).
- One would expect that cross-cutting relationships would be indicative, e.g. where a flute cast cuts a burrow. But, if the groove is deep and sharply bounded, then the producer could have cut straight through the groove, though this is unlikely (Fig. 8.22(e)).

And here are some criteria for distinguishing *post-depositional* traces. Care must be exercised where load casts have formed.

- Modification of the sand–mud interface by trace or detachment of the burrow from the sole (Fig. 8.22(f)).
- Trace conforms to shallow groove cast (Fig. 8.22(g)).
- Grooves deflected slightly by burrow.
- Cross-cutting relationship must be used with care.
- Large traces unaffected by current action.

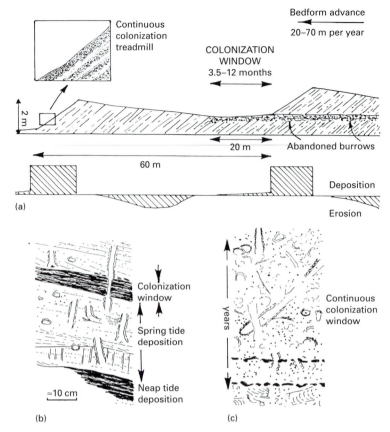

Fig. 8.23 Colonization window: (A) sand wave migration and opportunity for colonization of the trough area and dune foresets; (b) schematic to show depositional processes on an estuarine point bar where the opportunity for colonization is limited to periods of mud deposition; (c) shoreface environment where physical restraints on colonization are limited to infrequent storms. Pollard *et al.* (1993).

The tops of turbidites are often intensely bioturbated (Fig. 8.25, p. 157), especially in late Mesozoic and Cenozoic sediments, following the rapid increase in diversity of the post-event suite at the end of the Cretaceous, especially by echinoids. Search for distinct traces. If the study area is sufficiently large, then the turbidites and background sediment may show lateral changes in sedimentary character, proximal to distal (Fig. 8.26, p. 158). There may also be changes in the vertical sequence, e.g. associated with a coarsening-upward sequence.

In *storm and tempestite beds* there is generally less difference between the pre-event and post-event suites. Bioturbation may be pervasive, rendering the sharp sole indistinct, due to extensive and deep *Ophiomorpha* and *Thalassinoides*. Interbeds between events are also typically bioturbated. Again, be on the lookout for amalgamation. Most storm beds are compound. Incomplete trace fossils can give an indication of the amount of erosion that took place (Fig. 8.21).

Among *tidal and tidal-influenced facies*, high-energy cross-stratified sands and oolites seldom display extensive bioturbation. If bioturbation can be found, it is likely to be in two settings (Fig. 8.23):

- Along the 'bottom set', indicating colonization between large bedforms (dunes).
- Extending upwards along the cross-laminae (or muddy interlaminae), indicating pauses in the advance of the dune front.

Owing to the rapid changes that take place in high-energy tidal channels, keep a lookout for chance preservation of a marine trace fossil. This may be important in deciding whether one is dealing with a

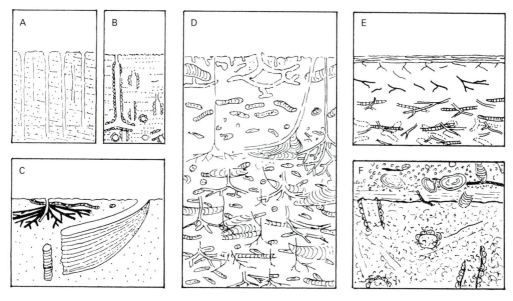

Fig. 8.24 Styles of tiering. (A) Single colonization produced by opportunists, e.g. *Skolithos*. (B) Single succes-sive colonization of *Ophiomorpha* on aggrading beach. (C) Simple tiering (frozen tier) as first colonization of an event bed. (D) Complex tiering associated with aggradation and overprinting in a deep-sea mud; for shelf sand see Fig. 8.8. (E) Gradational overlap reflecting gradual changes in ecological parameters, here schematically after Savrda and Bottjer (1989) with deoxygenation of the substrate; see also Fig. 5.6. (F) An omission overlap marks a sharp facies change to firmground at a key stratal surface.

fluvial or delta distributary environment. Only an ephemeral event, such as when a salt wedge has ex-tended into a distributary system, may be represented.

Deposition of an *ash* is often disastrous for the biota. Anticipate that there may be much biologically relevant information to be gleaned below the ash. With *airfall tuffs* search for epigenic traces: trails, faecal pellets, footprints, funnel openings. The ichno-logy and sedimentology of *marine tephras* have not yet been investigated. In shallow-marine sediments a sharp erosive base is usual (because the tsunami reached the site before the wind-blown ash cloud), and it is rare for any body fossils to be preserved autochthonously. Burrows were truncated. Because a large area was affected, expect that biotic recovery, if the top of the bed is preserved, may show evidence for gradual opportunistic recolonization (Fig. 8.30, p. 163).

8.5.5 The colonization window

The amount of bioturbation present in any facies must relate to the opportunities that were present for larval settlement or relocation. The period over which this was available, the *colonization window*, can be estim-ated for the primary sedimentation (Fig. 8.23). Other

indeterminate factors, especially ecological factors such as seasonality and temperature, would also have been important. Field estimation can be linked to the biological input, and Fig. 8.23 shows three examples.

Many trace fossils must have been formed rapidly. But *Diplocraterion* was able to adjust and extend over several depositional phases. In contrast, the com-plex networks of *Paleodictyon* or the complex spirals of *Zoophycos* (Figs 8.4 and 8.13) suggest that their construction took a substantial period of time. Their occurrence is always in relatively stable environments.

8.5.6 Tiering

Tiering above the substrate surface is discussed in Chapter 5 (Fig. 5.21). Below, the substrate is par-titioned into tiers by the infaunal biota in order to accommodate themselves (size) and in order to use different feeding and respiratory strategies. As aggradation takes place in an equilibrium situation, the tiers will move upwards so that lower tiers are superimposed on higher tiers, thus tending to obliter-ate the higher tiers and their information potential (Table 8.4 and Fig. 8.24).

Several illustrations of substrate tier reconstructions have been published. How these have been arrived at

Table 8.4 Tiering and ordering of ichnofabrics.

Observations	Implications	Environmental stability	Porosity/permeability (p/p)
Single colonization[a]	Opportunistic colonization by pioneers (low diversity, often, small size)	Generally associated with high-stress environment, e.g. beach, estuary	Enhancement of permeability normal to bedding common
Simple tiering	Colonization of an event bed by tiered suite	Typical of wide range of event beds	As above, or reduced p/p at bioturbated level
Single successive colonization[b]	Essentially gradual aggradation in low-diversity (stressed) environment	Relatively higher than above	Depends on type of trace
Complex tiering	Gradual aggradation of tiered suite	Stable environment	Reduced p/p
Gradational overlap	Gradual environmental change, e.g. in oxygenation		As above
Omission overlap	Sharp facies change associated with omission ± erosion to firmground or hardground; key stratal surfaces are the transgression surface and the sequence boundary		Generally associated with p/p break; occasionally 'open' maze/boxwork enhances p/p

[a] Primary structures generally clear.
[b] Repeated colonization ± aggradation.

is not always clear! Under more or less continuous sedimentation the depth of each bioturbator can be established only by working out the order in which each tier was introduced, based on cross-cutting relationships and careful examination of the associated primary structures, e.g. relationship with small-scale cross-lamination (Fig. 8.25). This is seldom sufficient to reconstruct a tier diagram with specific depths. An event bed may have frozen the tiering in the underlying sediment but erosion will often have removed shallow tiers (Fig. 8.21) and certainly rendered the mixed layer unrecognizable. Furthermore, the recolonizing biota will have a different make-up and tier structure, because a different substrate was present, as well as other ecological factors. The different styles of tiering closely relate to environmental parameters and are discussed further below.

To construct a tier diagram:

1 Observe as much of a face as time allows and note cross-cutting relationships (best done on a matrix diagram).
2 Photograph and dissect for further evidence.
3 Repeat for sample consistency.
4 Collect blocks with known orientation for laboratory slabbing (if practicable).

Ichnoguilds

The guild concept (groups of species that exploit a particular environmental resource) has proved useful in ecology and has been extended to palaeoecology

by Bromley (1996). As yet, few studies have been carried out. Potentially, ichnoguilds will form a useful basis for universal palaeoenvironmental interpretation and in the understanding of ichnofabrics.

Each guild may be defined using three categories:

• Spatial relationship, i.e. tier level.
• Dwelling or semipermanent burrows, or vagile, transitory burrows.
• Food source: deposit feeding or suspension feeding or farming (or chemosymbiosis).

For example, in the Upper Cretaceous Chalk of northwest Europe, and in deeper basinal settings, the *Chondrites–Zoophycos* ichnoguild can be recognized as deep-tier, non-vagile deposit feeding (or chemosymbiotic) structures. Besides *Chondrites* and *Zoophycos*, certain types of *Teichichnus* and also *Trichichnus* may be included. The traces represent deeply burrowing specialists.

8.5.7 Trace fossils and oxygenation

Highly bioturbated sediment can represent no other than a well-oxygenated seafloor. If finely laminated black mudrock represents the opposite (for other features see Section 5.3.6), then successions that show gradations, one into the other, represent deteriorating or ameliorating oxygen gradients (Fig. 5.6). Stratigraphic changes in oxygenation are readily recognized in the field, and especially in slabbed core. Changes

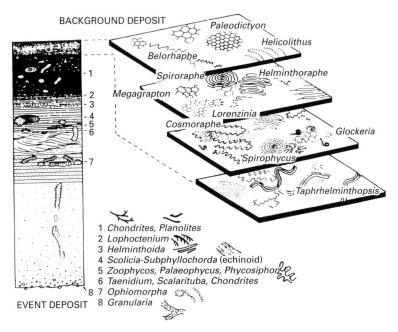

BACKGROUND DEPOSIT

Paleodictyon

Helicolithus

Belorhaphe

Spiroraphe

Helminthoraphe

Megagrapton

Lorenzinia

Cosmoraphe

Glockeria

'Spirophycus'

Taphrhelminthopsis

1 *Chondrites, Planolites*
2 *Lophoctenium*
3 *Helminthoida*
4 *Scolicia-Subphyllochorda* (echinoid)
5 *Zoophycos, Palaeophycus, Phycosiphon*
6 *Taenidium, Scalarituba, Chondrites*
7 *Ophiomorpha*
8 *Granularia*

EVENT DEPOSIT

Fig. 8.25 Generalized trace fossil tiering (frozen) within the turbidite (event) layer and complex tiering within the background sediment. Dotted lines indicate traces of older tiers. Based on the Upper Cretaceous–Lower Eocene of northern Spain. Tiering in the turbidite layer is controlled by the textural and structural profile. Tiering of the background sediment (carbonate poor) reflects the ranges of penetration depth, nutrient and oxygenation levels. More intensively bioturbated sediments are impoverished in graphoglyptids (e.g. *Paleodictyon*). Adapted from Leszcynski (1993).

seen in the rock have been shown to be close, if not identical, to changes observed in modern deep-sea cores. In response to decreasing oxygenation, the three principal changes are as follows:

- Decrease in trace size (burrow diameter).
- Decrease in trace diversity.
- Decrease in the number of tier levels, with loss of higher tiers, and associated with an upward shift of tiers.

If *Chondrites* is present, all three changes may be evident in this ichnotaxon, but it is safer not to rely on a single taxon. Shift of tier depth is obviously less well appreciated since it will be related to sedimentation rate. Gradual change probably took place over a considerable period of time, in contrast with an oxygenation 'event' or seasonal change. Besides oxygenation, two other ecological factors should also be kept in mind: sedimentation rate and nutrient supply, especially the availability of suspended food (Fig. 8.26). As the nutrient supply decreases, animals

will have to forage for food. It is now recognized that nutrients are not uniformly available in the deep sea. Sedimentation rate can be estimated from the lamination and context. When the sedimentation rate is low but the input of organic matter high, the sediment is typically brown to green in colour. The evidence is especialy clear in the frozen tiers at the base of turbidites where minimal erosion has occurred (Wetzel and Uchman, 1999).

At outcrop one has the advantage of being able to examine a far greater lateral extent and/or bedding surfaces than in core, when it is necessary to place particular emphasis on the ichnology.

8.5.8 Salinity

The ichnology of brackish-water facies is somewhat controversial. At present we do not understand sufficient about the biological controls: particularly how marine animals cope with reduced salinity, both within species and between species and groups. In

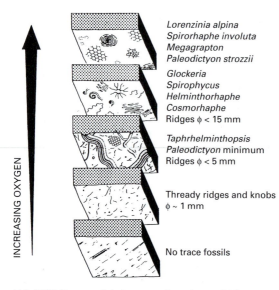

INCREASING OXYGEN

Lorenzinia alpina
Spirorhaphe involuta
Megagrapton
Paleodictyon strozzii

Glockeria
Spirophycus
Helminthorhaphe
Cosmorhaphe
Ridges φ < 15 mm

Taphrhelminthopsis
Paleodictyon minimum
Ridges φ < 5 mm

Thready ridges and knobs
φ ~ 1 mm

No trace fossils

Fig. 8.26 Oxygen-related succession of preturbidite trace fossils seen on turbidite soles (Cretaceous to Tertiary). The upper diagram represents optimum conditions for graphoglyptid producers (e.g. *Paleodictyon*). Recent work by Wetzel and Uchman (1998) indicates that sedimentation rate may be the significant factor, and that graphoglyptids may be characteristic of low nutrient supply. Adapted from Leszczynski (1991).

tidal environments deep burrowers find temporary refuge and can penetrate further upstream than epifaunal animals. It is clear that some animals (bivalves, e.g. *Mytilus* in the Baltic) become stunted in reduced salinities, but this does not apply to the annelid traces. Some arthropods indeed are able to adapt physiologically without any decrease in size. But since we can rarely identify the culprit of a trace in ancient sediments, we have no means of telling whether its constructor could adapt.

A number of criteria have been taken to indicate brackishness in ichnotaxa. There is no disputing that there is a general reduction in diversity from open-marine conditions towards brackish conditions. But this is generally associated with high population densities, a feature of opportunistic colonization. The food resources of estuarine environments and lagoons are generally high. Some claim that there is a concomitant decrease in size, but this is not well established. Organism size (and hence burrow diameter) depends on many factors, including food availability, age and occupation time. There are no ichnotaxa that are specific to brackish-water facies. But if an integrated approach is adopted, with analysis of the facies and

palynofacies and evaluating the *colonization window*, then a useful environmental interpretation can be made.

In lagoonal sediments (Box 6.3) a suite of shells is generally present, and 'salinity events' may be indicated by a single occurrence of *Diplocraterion* (e.g. Fig. 8.28, p. 161), representing a flooding surface in Mesozoic sediments. In estuaries a salinity gradient is to be expected, ranging from normal salinity to freshwater. Taking an integrated approach is all the more important.

8.5.9 Diversity

It is always tempting to exaggerate the number of ichnotaxa present in a facies. This can be done all too easily by applying available names. But ask questions such as, Are the vertical burrows (e.g. *Skolithos*) merely the shafts of a complex system at depth which have been identified as *Zoophycos*? Other more subtle examples may be cited. Differences in preservation have been cited as distinct taxa.

8.5.10 Trace fossils and palaeocommunities

This chapter has perhaps overemphasized the role of trace fossils in facies interpretation. But how can the traces contribute to the recognition of palaeocommunities (Section 5.4)? With care, they can be included as an indication of the number, diversity and structure of the soft-bodied component. But for each case:

- The trace fossil analysis must be linked to the body fossil analysis, either as part of a common autochthonous assemblage or, indirectly, to a time-averaged assemblage.
- The tiering structure and ordering must be analysed, especially to distinguish between trace fossil assemblages of different colonizations, e.g. pre-event and post-event suites.
- Care needs to be taken in measuring the diversity.

Alternatively, where body fossils are absent, a trace fossil assemblage may be interpreted as an ichnocoenosis in its own right. This is particularly useful for assemblages of vertebrate footprints (Lockley, 1991; Lockley *et al.*, 1994), and for an appreciation of the palaeoecology of the diverse Permian arthropod ichnofauna of the Robledo Mountains, New Mexico (recorded in papers published by the New Mexico Museum of Natural History and Science).

8.6   Trace fossils and biostratigraphy

In general, trace fossil taxa are long-ranging. This is also a general prediction since they represent patterns of behaviour, each mode and pattern having been practised by different groups of organisms over much of geological time as different groups have occupied a particular niche. There are three exceptions:

1 Vertebrate feet are often very distinctive. One is reminded of the fascinating deductions made by North American Indians from footprints, not only the tribal identity by recognizing different moccasins and snowshoe design, but also the way in which the quarry was moving and its speed or possible injury. For the ichnologist, footprints offer information on the identity and ecology of the tracemakers. In continental Permo-Triassic deposits, where skeletal materials are rare to absent and palynological material often poorly preserved, a useful footprint stratigraphy has been developed.
2 The appendages of arthropods, especially many trilobites, evolved limbs for digging and burrowing. Different trilobite taxa evolved slightly different strategies, and the marks made also partly reflect differences in the morphology of the exoskeletal shield. For convenience they are referred to separate ichnospecies. Seilacher (1970, 1992) developed a *Cruziana* stratigraphy (Fig. 8.12) which is of particular value in Lower Palaeozoic facies where body fossils are poorly represented. Few body fossils have been found either actually with the trace, or as discrete fossils in associated strata. To prove the latter, it must be shown that the stratigraphical and geographical ranges of the ichnotaxon are identical to that of the inferred maker.
3 Some trace fossils, as presently understood, have a restricted stratigraphical range. In part this can be attributed to the evolution of a particular behavioural strategy.

Trace fossils at the Precambrian–Cambrian boundary

At the present time the Precambrian–Cambrian boundary is taken at the appearance of the deposit-feeding trace *Phycodes pedum*. This is a provisional decision: there must have been stages in the behavioural evolution of this quite complex trace. There are two aspects: ichnological changes during the late Proterozoic, and changes over the Precambrian–Cambrian boundary interval.

Although skeletonization took place rapidly (<10 m.y.) at around the Precambrian-Cambrian transition, metazoan origins (as evidenced by trace fossils) go much further back to c.1000 Ma. Interpretation of the late Proterozoic (Ediacaran) biota – long considered to represent extant animal groups (e.g. arthropods, echinoderms, coelenterates) – is now better regarded (though not by all workers) as evidence of an extinct clade (Vendobionta) with a make-up quite different from any living forms. A few survived into the early Palaeozoic.

A few of the traces in the late Proterozoic are known only from that stratigraphic interval. This is unexpected! Other traces that appear at about the boundary interval in shallow-water sediments seem to have 'migrated' to deeper-water sediments in younger rocks. This can be explained either by a migration to deeper water, or a subsequent restriction to deeper-water sediments where the peculiar organic mats that pertained in the late Proterozoic persisted.

For high-resolution ichnostratigraphy see Fig. 7.4.

8.7   Vertebrate tracks

Vertebrate footprints and trackways can provide much information about the ecology of the trackmaker. In describing vertebrate trackways, map the distribution of footprints and casts as accurately as possible and make the measurements indicated on Fig. 8.27. The morphology of the prints will depend on several factors (Section 8.1). Lockley (1991) views dinosaur tracks in a wide context and, for any site:

- Footprints may be the only evidence of dinosaurs and/or other large vertebrates.
- Footprints *may or may not* complement known skeletal remains.
- Footprints *may or may not* be present in the same proportion as skeletal remains.
- Footprints constitute only a small part of the available evidence about a taxon, and which *may or may not* be consistent with the skeletal evidence.
- The footprints of a (skeletal) taxon are unknown. This poses the question, Where might they be found?

The speed of travel represented by tracks can be estimated from the length of stride and foot length. Walking is indicated if the step (pace) length is less than four times the foot length, and running if the step (pace) length is greater than four times the foot length.

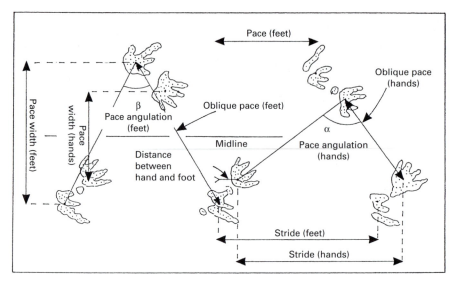

Fig. 8.27 Measurements for vertebrate trackway analysis. Adapted from Leonardi (1987).

8.8   Coprolites and faecal pellets

Coprolites and faecal pellets are common, especially as burrow fillings. But they may also be seen on bedding surfaces, where they can give an extended occurrence to a burrow, if burrow and pellet can be correlated; this may be the case when identifying a candidate *Diplocraterion* level in core (probably indicative of a flooding surface) without evidence of the actual burrow. The pellets often associated with Mesozoic *Thalassinoides* display a characteristic V-shaped sculpture externally (visible with a hand lens), and ordered tubules (seen in cross-section) indicative of their arthropod origin. Fossil leaves and wood often show evidence of insect feeding and other activity (Scott, 1992).

The larger coprolites (generally phosphatic) can be distinguished from pebbles by their resemblance to familiar faecal shapes. Their composition may give useful clues about diet (Hunt *et al.*, 1994).

8.9   Trace fossils and facies interpretation

The environmental (or Seilacherian ichnofacies) classification of trace fossils is the one most widely known and referred to, though often misleadingly. In the 1950s – when sedimentology was in its infancy and 'graded' muddy sandstones had only just been recognized as 'deep-water' in origin and the product of turbidity currents – Seilacher recognized the immense distinction between turbidite-associated trace fossil assemblages, and assemblages associated with shallow-marine sediments. He also recognized that these distinctions were related to behavioural aspects with a diverse suite of complex feeding traces dominating the deep-sea ichnobiota, and a suite often dominated by a suspension-feeding biota in shallower-marine environments. This bipartite distinction was later expanded to the well-known ichnofacies concept, each with its eponymous ichnotaxon. Since the scheme is well entrenched in the literature, a summary sketch is included (Fig. 8.18).

Although a few additional ichnofacies have been recognized, the resolution of the ichnofacies in respect of facies interpretation is quite insufficient for modern palaeonenvironmental studies, particularly in industry, and does not readily relate to allostratigraphic (sequence stratigraphic) correlation or key stratal surfaces. The ichnofacies concept is also difficult to apply to slabbed core, where an integrated approach between sedimentology and ichnology is essential. Although Seilacher recognized the essential link between the ichnology and sedimentary facies, this has been missing in much of the more recent work. However, what is still absolutely valid is the clear distinction between the ichnology of (1) *neritic*

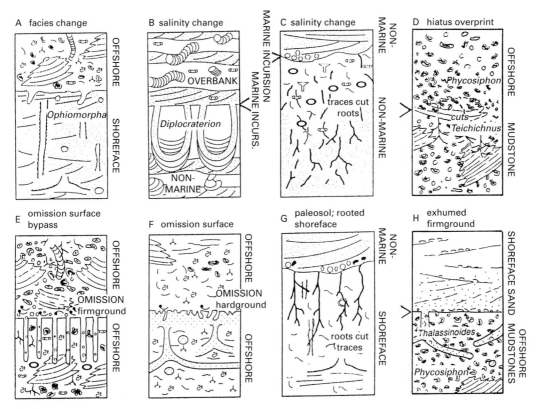

Fig. 8.28 How trace fossils can indicate key stratal surfaces, with examples from the Mesozoic of the North Sea and Yorkshire coast. (A) A marine flooding surface where offshore muds with *Teichichnus, Phycosiphon* and *Planolites* are transgressive over shoreface sands with scattered *Ophiomorpha*. (B, C) Marine incursions into non-marine sediments, where it is only the burrows that provide the evidence: (B) with *Diplocraterion*, and (C) where *Ophiomorpha* and other marine taxa cut roots; *Beaconites* pervades the non-marine sediments. (D) Subtle effects of a transgressive event in offshore mudstone facies with *Phycosiphon* (shallow tier) cutting *Teichichnus* (deep tier). (E, F) The effect of omission on colonization: (E) firmground *Skolithos* cross-cut offshore muddy sand with *Teichichnus*, followed by a similar offshore ichnofabric (omission indicates bypassing); (F) cementation of the omission surface has led to a bored hardground. (G) An immature soil profile cuts shoreface sands with *Ophiomorpha*, and is overlain by non-marine sands (omission with palaeosol). (H) Here omission has led to a firmground colonized by *Thalassinoides* cutting softground muds with *Phycosiphon*; shoreface sands follow. Adapted from Taylor and Gawthorpe (1993).

facies, (2) *deep-water muddy* facies (marine and lacustrine), (3) *non-marine* fluvial and floodplain facies, and (4) *event beds*, in respect of the trace fossil spectrum and styles of colonization and tiering in each (Figs 8.18 and 8.20). Distinguishing which of these four broad facies is present is a first step in facies interpretation (Table 8.2, panel C).

Recognizing *key stratal surfaces* (sequence boundaries, transgression surfaces, formation boundaries, unconformities, flooding surfaces, etc.) is the first objective in establishing sequences during the mapping and logging of cores and outcrops. Omission surfaces, firmgrounds and hardgrounds are associated

with distinctive styles of colonization and tiering, regardless of age. This establishes a framework. Several types of key stratal surface are shown in Fig. 8.28, including these:

- Surfaces that mark *omission without change* in substrate state or significant change in water depth (Fig. 8.28(D)).
- Surfaces that mark *omission with change* in substrate state and a significant change in depth (energy state) but *without immersion* (Fig. 8.28(E) and (H)). Such surfaces, marked by vertical or sharp-margined, passively filled burrows, are typical

of sequence boundaries or combined sequence boundary/transgression surfaces.

- Surfaces that mark *omission and emersion* (Fig. 8.28(G)), as at a sequence boundary where a substantial fall in relative sea level has taken place,
- Surfaces that mark a *flooding event* (± omission) within a non-marine succession (Fig. 8.28(B) and (C)), or from shoreface to deep-marine facies (Fig. 8.28(A)).
- Surfaces that mark *omission* (possibly through sediment bypassing) and *marked change in substrate state* (Fig. 8.28(F) and (H)) but with *little change in depth*.

Between key stratal surfaces the facies will be essentially conformable and either show little or no change, or display environmental shift, as in upward coarsening (shallowing) parasequences, representing change from offshore muds to shoreface sands (Fig. 8.29). Analysis of such *genetically related successions* provides the details of facies interpretation and basin development. Changes in the ichnofabrics may be gradual or abrupt, but there may be changes in colonization and tiering that can be correlated with the zonation of the shoreface and offshore depositional environments, as shown in Figs 8.8 and 8.29. Other genetically related successions have been mentioned in this text, and here are some of the trends they demonstrate:

- Proximal–distal trends of turbidites and storm beds (Fig. 6.29).
- Oxygenation trends in black shales (Fig. 5.7) and turbiditic facies.
- Salinity fluctuations in marginal marine (lagoonal) facies.
- Depth-related lateral changes in firmgrounds and hardgrounds.

The three groups of criteria based on sedimentology, colonization and tiering style, form a framework for the ichnological interpretation of facies which can be integrated into allostratigraphic (sequence stratigraphic) schemes. Fig. 8.30 shows four ichnological curiosities, which reveal quite significant information.

8.9.1 Methodology

As with a suite of body fossils, there are two principal stages to analysis: autecological and synecological. The alternative – to pick certain traces and attempt to slot them into described ichnofacies and environmental models – is unscientific. Pose the question,

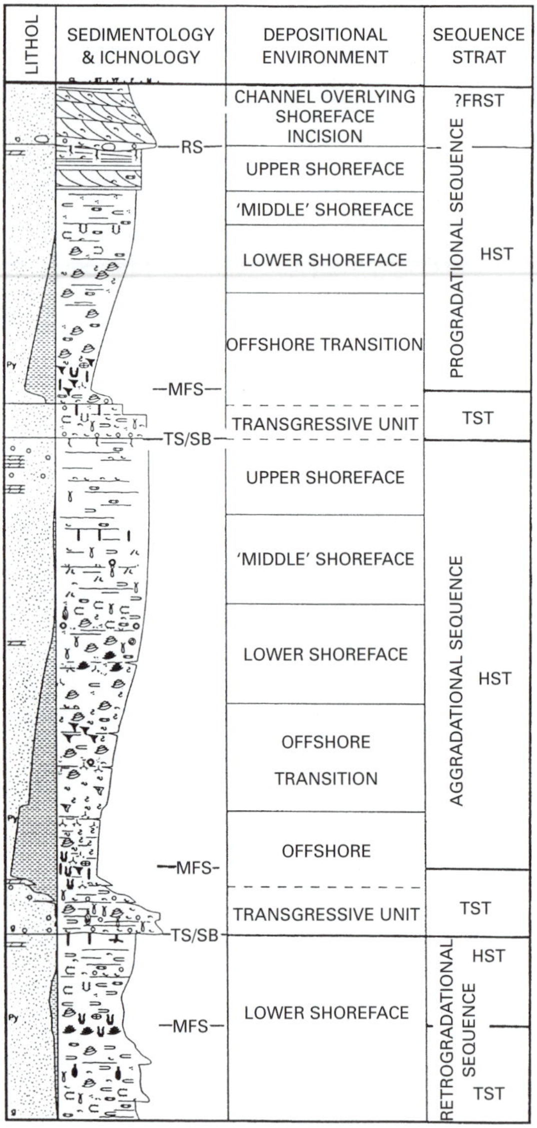

Fig. 8.29 Core log to illustrate a genetically related succession based on a Jurassic North Sea core. It illustrates successive ichnofabrics with depositional environments as defined by Ichron Ltd (see Fig. 8.8) Environmental boundaries will vary slightly depending on individual interpretations. Systems tracts (sequence stratigraphy): TS/SB = transgression surface/sequence boundary, MFS = maximum flooding surface, FRST = forced regression sequence tract, HST = highstand systems tract, TST = transgressive systems tract.

What colonization and tiering styles, and sedimentary structures might be expected in the environment being postulated? In practice, five approaches are used to analyse and interpret trace fossils within facies:

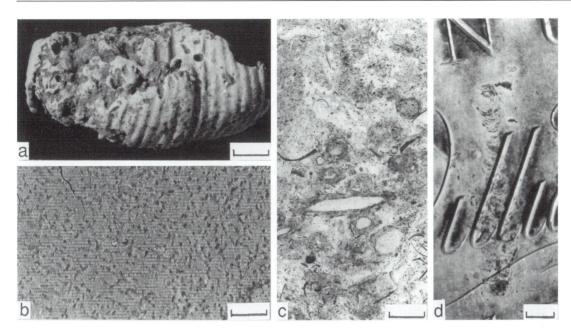

Fig. 8.30 Four ichnological curiosities of some significance. (a) Part of an infilled ammonite (body chamber) reworked from Upper Jurassic limestones into Lower Cretaceous and worn and extensively bored; Box 6.2. (b) Dense, small vertical burrows in laminated bentonitic clay and silt representing the first stage of recolonization following a major tephra event in the Lower Cretaceous of southern England (Goldring, 1996). (c) Reworked and thick burrow linings (*Diopatrichnus*) in marginal marine facies of Middle Jurassic of central England (de Gibert, 1996). (d) *Teichichnus rectus* on gravestone in Ratby churchyard, central England, quarried from Cambrian slates, long regarded as Precambrian, till found by amateur geologist Ben Bland. Scale bars = 10 mm.

Using a standard bathymetric profile The ichnotaxa are identified and the list is compared with the distributions on a standard bathymetric profile, generated by experience (Chamberlain, 1971). This method is analogous to traditional biostratigraphic procedure. The problem is that the same ichnotaxa can occur in widely different environments.

Grouping ichnotaxa into associations Ichnotaxa are grouped into associations (ichnocoenoses) in order to recognize trophic and ethologic positions. For instance, working in the Lower Jurassic of East Greenland, Dam (1990) found there was a strong correlation with the sedimentary environments that he had recognized from the primary sedimentary structures. The problem here is that one can easily overlook the importance of colonization surfaces, especially key stratal surfaces, when relating the ichnotaxa to the host sediment. In the field one tends to focus on extensive bedding surfaces with interstratal traces, and pay less attention to highly bioturbated sediment, from which it is difficult to collect readily and/or identify the ichnotaxa.

Using the Seilacherian ichnofacies scheme In addition to the problem of resolution, the scheme does not satisfactorily deal with deltaic or estuarine (delta distributary) facies and environments. Also many of the diverse ichnotaxa seen at outcrop on bedding surfaces are notoriously difficult or impossible to identify in core in two dimensions, e.g. *Cruziana, Paleodictyon*.

Integrating sedimentology and ichnotaxa In recent years there has been an emphasis on ichnofabrics, which integrate the sedimentology and the various ichnotaxa (and their ordering). The integrated approach yields greater resolution (Taylor and Gawthorpe, 1993; Martin and Pollard, 1996). The studies have been made with respect to successions synthesized from a number of North Sea wells in Upper Jurassic sandstones. However, it is clear that different ichnofabrics (with different ichnotaxa) will represent different successions of different geological ages, and probably climatic zones. Bromley suggested that overall there might be perhaps a hundred or so different ichnofabrics, but this is very much a guess.

No formal scheme has been used to name the various ichnofabrics, and in many respects using ichnofabrics involves much pattern recognition.

Integrating sedimentology, colonization and tiering   The method outlined earlier on is process-related and places emphasis on styles of colonization and tiering that are independent of ichnotaxa and geological age. But it is always useful to identify the taxa for their palaeoecological and geotechnical roles. This method has been found to work well in the field and core shed, and on sediments of various ages.

8.9.2   Summary of ichnofabrics and tiering in siliciclastic facies

Tucker (1996, Ch. 8 and Tables 8.2–8.12) gives a useful synthesis of sedimentary facies and environments, and sequence stratigraphy for use in the field.

The shore–offshore profile   The intensity of bioturbation depends on sedimentation rate (Table 7.4).

- *Offshore*: where the bioturbation index (BI) is high, thin event beds will be scarcely recognizable; complex tiering is normal, but with irregular lateral and vertical changes in ichnotaxa; the maximum flooding surface has a highly bioturbated or restricted muddy interval (Section 6.5).
- *Offshore–shoreface transition*: any storm beds display rapid recolonization to background levels, though with local opportunistic colonization and simple tiering; tempestites associated with more pronounced opportunism; muddy storm units are generally not colonized.
- *Shoreface with composite storm beds*: this displays parallel to bioturbated succession ± erosional tops and recolonization; from the Mesozoic onwards there is complete overprinting by deep burrowers (including bivalves) and primary lamination is minimal; with shallowing, discrete mud layers become uncommon, but evidence for reworking increases, along with the proportion of shafts to galleries, and lined to unlined burrows; shore sands indicated by single or single successive colonization.

Marginal marine   In a delta distributary or estuary, the channel sediments exhibit minimal bioturbation. There are point-bar heterolithics with sporadic single colonizations (or simple tiering) from mud drapes into inclined sand sheets. Dense *Palaeophycus* or *Macaronichnus* are good indicators of (near) normal salinity. Burrows may be relatively incomplete.

Locally, sandwaves with foresets are intensely burrowed by a single ichnotaxon; troughs are intensely bioturbated and the diversity reflects any salinity decrease. Deeper-burrowing taxa penetrate more or less normal to the lamination. Ancient intertidal ichnology is scarcely known.

Lagoonal ichnofabrics   Lagoonal ichnofabrics reflect brackishness. Salinity events are indicated by opportunistic single colonizations. General bioturbation reflects salinity and oxygenation fluctuations, but the lower levels of penecontemporaneous erosion and generally finer grain size allow preservation of epigenic and very shallow traces. Ostracode and bivalve resting traces (± adjustment) and locomotion tracks occur. Spillover sands poorly bioturbated with more or less rapid return to lagoonal muds. Presence of *Palaeophycus* (and possibly other marine ichnotaxa) distinguishes lagoonal or delta embayment facies from delta plain lakes.

Deep-water muds   Fluctuations in sedimentation rate, nutrient supply and oxygenation are reflected in the tier structure and tier depths (Fig. 5.6).

Turbidites   Distal turbidites may cover a 'frozen' tier profile. Recolonization reflects any change in ecological parameters. Proximality is reflected in pronounced simple tiering ± relocated burrowers.

Non-marine facies   In general there is single or simple tiering (± intense bioturbation). There are epigenic trails in shallow water (ponds, lakes subject to desiccation). Channel deposits are not colonized, or are colonized only following channel switching or at seasonal low-flow conditions, and then the colonization is by floodplain biota with single (single successive or simple tiering) ± adjustment structures, by bivalves, and frequently overprinted by roots (with succesive colonizations interrupted by burrowing during wetter periods). Delta plain facies are characterized by single colonizations (opportunisitc), rooted horizons and plant accumulation, and often laterally extensive minor flooding events (again with single colonizations). Deep lakes seem to have a similar but less diverse ichnology to deep-marine muds.

8.9.3   Carbonate facies ichnology

It may seem illogical to separate carbonate sediments from siliciclastic sediments when discussing their ichnology, but this is justified by the differences in their lithification potential and diagenetic history (Section 2.8). The following points can be made:

1 Heterolithic micrite–grainstone facies show similar ichnologies to sand–mud facies. This also applies, in general, to event beds in carbonates, e.g. storm units or calcareous turbidites. In calcareous turbidites the traces within the bed are clearer than in siliciclastic turbidites.

2 In marginal marine facies the bedding is typically compound and amalgamated. Slabbing and analysis of the ichnofabrics can reveal the nature of the primary stratification (de Gibert, 1996).

3 Skeletal elements inhibit burrowing, but are susceptible to boring and encrustation.

4 The general make-up of carbonate sediment is generally the best indication of depositional environment.

5 Carbonate sediments, especially grainstones, have a high lithification potential. This may accentuate trace fossil definition, but in fine-grained carbonates it may inhibit lithification.

6 Firmgrounds, and especially hardgrounds, are common and important stratigraphically and in permeability studies. They were formed rapidly.

7 Following subaerial exposure, rhizomorphs are widespread and of stratigraphical significance.

8 Fine-grained carbonates do not compact in the same way or extent as siliciclastic muds, but there is the problem of dissolution and stylolite formation, with 'underbeds' and 'overbeds' (Fig. 2.2), and this obscures bedding and interface burrows. Dissolution can also lead to a nodularity, which simulates burrows; or the burrow may become a site for cementation.

9 Resedimented chalks have their primary porosity between the clasts. The ichnology of the clasts provides clues to provenance.

10 There are five main types of chalk: shelf-sea chalks with flint-filled or chert-filled *Thalassinoides*, together with *Chondrites*, *Planolites* and *Zoophycos*; hardgrounds, which may occur in groups; winnowed chalks, which are devoid of trace fossils; deep-sea chalks with *Chondrites*, *Planolites* and *Zoophycos*; allochthonous chalk.

8.10 Trace fossil description

Description of trace fossils is generally necessary in the field since many are too large or too difficult to extract (see also Section 2.8 and Fig. B.3, p. 177).

1 Note location and position in the bed.

2 Ascertain the stratigraphy and facies.

3 Attempt to ascertain the grade of bioturbation (bioturbation index) and frequency of each ichnotaxon for specific facies (above).

4 Try to understand the origin (epigenic, endogenic) and mode of preservation of each trace.

5 For bedding-parallel traces, establish whether they are epigenic or endogenic, and whether any pass into oblique to vertical traces, or have an oblique to vertical component. Begin each description with overall form and dimensions, then details: straight, sinuous, meandering, branching, form in cross-section and sculpture, distribution and form of prints, dimensions and spacings.

6 For all burrows describe their attitude to the bedding (parallel, oblique, vertical); orientation, preferred orientation, spacing, then go on to details of form, e.g. vertical, U-form, branching ± spreite, nature of termination(s), nature of burrow fill (meniscate, symmetrical or asymmetrical, annular, unstructured, pelleted) and manner of filling (more difficult to determine) (passively by sedimentation with complete or partial collapse and deformation, actively by organism, or complete collapse), and relationship to adjacent sediment; nature of burrow margin and whether margin scratched; nature of burrow lining (if any), its composition and thickness. Make a general assessment of the relative abundance of vertical and horizontal burrows.

7 Attempt to determine the basic form of the burrow occupied by the animal and then the manner in which it was extended.

8 Attempt to establish the relative timing of formation of each type of trace (ordering) and the tiering. Both objectives can be difficult and time-consuming. Tiering is clearest below the upper surface of event beds in the post-event suite, but check for 'frozen' tiers below the base. Traces that are normally in the shallow tier (e.g. *Rhizocorallium, Planolites, Palaeophycus*) can extend more deeply in reworking the fill of other burrows (e.g. *Planolites* in *Thalassinoides*). Ordering (the sequence of burrows) is clearest at omission surfaces and is generally best analysed on bedding surfaces. Attempt a portrait (icon) of the overall ichnofabric, and how it was produced.

9 Determine any vertical change in degree of bioturbation, burrow types and in diversity. Slabbing may provide more information. Consider carefully before cutting, and try to anticipate what may be

revealed. For traces normal to bedding, slabbing should be parallel to bedding. Thus, to distinguish a meniscate burrow as a retrusive *Diplocraterion* and not *Teichichnus*, the cut should be made perpendicular to the direction of retrusion. In ichnofabric analysis it is important to appreciate (and describe and figure) sections both normal and parallel to the stratification (or lamination).

10 Determine the diversity.

11 Sketch and photograph.

12 When collecting, take care to sample any variation that may be due to behavioural differences. Where applicable, always collect counterpart material for stratinomic analysis.

References

Bromley, R. G. (1991) *Zoophycos*: strip mine, refuse dump, cache or sewage farm? *Lethaia* **24**: 460–62

Bromley, R. G. (1994) The palaeoecology of bioerosion. In Donovan, S. K. (ed.) *The palaeobiology of trace fossils*, Johns Hopkins University Press, Baltimore MD, pp. 134–54

Bromley, R. G. (1996) *Trace fossils: biology, taphonomy and applications*, 2nd edn. Chapman & Hall, London (A very readable text with good appreciation of modern traces and the palaeobiology of ancient traces; rather thin sedimentologically and does not discuss hardgrounds; includes a glossary)

Chamberlain, C. K. (1971) Morphology and ethology of trace fossils from the Ouachita Mountains, southeast Oklahoma. *Journal of Paleontology* **45**: 212–46

Dam, G. (1990) Palaeoenvironmental significance of trace fossils from the shallow marine Lower Jurassic Neill Klinter Formation, East Greenland. *Palaeogeography, Palaeoclimatology, Palaeoecology* **79**: 221–48

de Gibert, J. M. (1996) *Diopatrichnus odlingi* n. isp. (annelid tube) and associated ichnofacies in the White Limestone (Middle Jurassic) of Oxfordshire: sedimentological and palaeoecological significance. *Proceedings of the Geologists' Association*, **107**: 189–98

Droser, M. L. and Bottjer, D. J. (1987) Development of ichnofabric indices for strata deposited in high-energy nearshore terrigenous clastic environments. In Bottjer, D. J. (ed.) *New concepts in the use of biogenic sedimentary structures for paleoenvironmental interpretation*. Society of Economic Paleontologists and Mineralogists, Pacific Section, Los Angeles, pp. 29–33

Droser, M. L. and Bottjer, D. J. (1989) Ichnofabric of sandstones deposited in high-energy nearshore environments: measurement and utilization. *Palaios* **4**: 598–604

Goldring, R. (1996) The sedimentological significance of concentrically laminated burrows from Lower Cretaceous Ca-bentonites, Oxfordshire. *Journal of the Geological Society, London* **153**: 255–63

Goldring, R., Astin, T. R., Marshall, J. E. A., Gabbott, S. and Jenkins, C. D. (1998) Towards an integrated study of the depositional environment of the Bencliff Grit (Upper Jurassic) of Dorset. In Underhill, J. R. (ed.) *Development and evolution of the Wessex Basin*. Special Publication 133, Geological Society, London, pp. 355–72

Gruszczynski, M. (1986) Hardgrounds and ecological succession in the light of early diagenesis: Jurassic, Holy Cross Mountains, Poland. *Acta Palaeontologica Polonica* **31**: 163–212

Hunt, A. P., Chin, K. and Lockley, M. G. (1994) The palaeobiology of vertebrate coprolites. In Donovan, S. K. (ed.) *The palaeobiology of trace fossils*, Johns Hopkins University Press, Baltimore MD, pp. 221–40

Jensen, S. (1997) Trace fossils from the Lower Cambrian Mickwitzia Sandstone, south-central Sweden. *Fossils and Strata* **42**: 1–111

Kern, J. P. (1980) Origin of trace fossils in Polish Carpathian flysch. *Lethaia* **13**: 347–62

Leonardi, G. (ed.) (1987) *Glossary and manual of tetrapod footprint palaeoichnology*. Departamento Nacional da Produção Mineral, Brasilia

Leszczynski, S. (1991) Oxygen-related controls on predepositional ichnofacies in turbidites, Guipuzcoan Flysch (Albian–Lower Eocene), northern Spain. *Palaios* **6**: 271–80

Leszczynski, S. (1993) A generalised model for the development of ichnocoenoses in flysch deposits. *Ichnos* **2**: 137–46

Lockley, M. (1991) *Tracking dinosaurs*. Cambridge University Press, Cambridge

Lockley, M. G., Hunt, A. P. and Meyer, C. A. (1994) Vertebrate tracks and the ichnofacies concept: implications for palaeoecology and palichnostratigraphy. In Donovan, S. K. (ed.) *The palaeobiology of trace fossils*, Johns Hopkins University Press, Baltimore MD, pp. 240–68

MacNaughton, R. B. and Pickerill, R. K. (1995) Invertebrate ichnology of the nonmarine Lepreau Formation (Triassic), southern New Brunswick, eastern Canada. *Journal of Paleontology* **69**: 160–71

Martin, M. A. and Pollard, J. E. (1996) The role of trace fossil (ichnofabric) analysis in the development of depositional models for the Upper Jurassic Fulmar Formation of the Kittiwake Field (Quadrant 21 UKCS). In Hurst, A. *et al., Geology of the Humber Group: Central Graben and Moray Firth, UKCS*, Special Publication 114, Geological Society, London, pp. 163–83

Palmer, T. J. and Palmer, C. D. (1977) Faunal distribution and colonization strategy in a Middle Ordovician hardground community. *Lethaia* **10**: 179–99

Perry, C. T. (1998) Grain susceptibility to the effects of microborings: implications for the preservation of skeletal carbonates. *Sedimentology* **45**: 39–51

Pollard, J. E., Goldring, R. and Buck, S. G. (1993) Ichnofabrics containing *Ophiomorpha*: significance in shallow-water facies interpretation. *Journal of the Geological Society, London,* **150**: 149–64

Savrda, C. E. and Bottjer, D. J. (1989) Trace-fossil model for reconstructing oxygenation histories of ancient marine bottom waters: application to Upper Cretaceous Niobrara Formation, Colorado, *Palaeogeography, Palaeoclimatology, Palaeoecology* **74**: 49–74

Scott, A. (1992) Trace fossils of plant–arthropod interactions. In Maples, C. G. and West, R. R. (eds) *Trace fossils.* Short Courses in Paleontology 5, Paleontological Society, pp. 197–223 (The society has a publications office at the Carnegie Museum of Natural History, Pittsburgh PA)

Seilacher, A. (1964) Biogenic sedimentary structures. In Imbrie, J. and Newell, N. (eds) *Approaches to paleoecology.* J. Wiley, New York, pp. 296–316

Seilacher, A. (1970) *Cruziana* stratigraphy of 'non-fossiliferous' Palaeozoic sandstones. In Crimes, P. T. and Harper, J. C. (eds) *Trace fossils.* Special Issue 3, *Geological Journal,* pp. 447–76

Seilacher, A. (1992) An updated *Cruziana* stratigraphy of Gondwanan Palaeozoic sandstones. In Salem, M. J. and Busrewil, M. T. (eds) *The geology of Libya.* Elsevier, Amsterdam, pp. 1565–81

Taylor, A. M. and Gawthorpe, R. L. (1993) Applications of sequence stratigraphy and trace fossil analysis to reservoir description: examples from the Jurassic of the North Sea. In Parker, J. R. (ed.) *Petroleum Geology of Northwest Europe: Proceedings of the 4th Conference.* Geological Society, London, pp. 317–35

Taylor, A. M. and Goldring, R. (1993) Description and analysis of bioturbation and ichnofabric. *Journal of the Geological Society, London,* **150**: 141–48.

Tucker, M. E. (1996) *Sedimentary rocks in the field.* J. Wiley, Chichester

Wetzel, A. (1984) Bioturbation in deep-sea fine-grained sediments: influence of sediment texture, turbidite frequency and rates of environmental change. In Stow, D. A. V. and Piper, D. J. W. *Fine-grained sediments:*

deep water processes and facies, Geological Society, London, pp. 595–608

Wetzel, A. and Aigner, T. (1985) Stratigraphic completeness: tiered trace fossils provide a measuring stick. *Geology* **14**: 234–37

Wetzel, A. and Uchman, A. (1998) Deep-sea benthic food context recorded by ichnofabrics: a conceptual model based on observations from Paleogene Flysch, Carpathians, Poland. *Palaios* **13**: 533–46

Wilson, M. A. and Palmer, T. J. (1992) *Hardgrounds and hardground faunas.* Publication 9, University of Wales, Aberystwyth, Institute of Earth Studies, pp. 1–131

Further reading

Collinson, J. D. and Thompson, D. B. (1988) *Sedimentary structures.* 2nd edn. Allen & Unwin, London (Contains a useful section on biogenic sedimentary structures)

Donovan, S. K. (ed.) (1994) *The palaeobiology of trace fossils.* Johns Hopkins University Press, Baltimore MD

Ekdale, A. A., Pemberton, S. G. and Bromley, R. G. (1984) *Trace fossils.* Short Course Notes 15, Society of Economic Paleontologists and Mineralogists, Tulsa OK (A useful but dated text for sedimentological applications)

Pemberton, S. G. (ed.) (1992) *Applications of ichnology to petroleum exploration.* Core Workshop Guide 17, Society of Economic Paleontologists and Mineralogists, Tulsa OK (A collection of papers mostly relating to the Mesozoic of Alberta)

Pemberton, S. G., Maceachern, J. A. and Frey, R. W. (1992) Trace fossil facies models: environmental and allostratigraphic significance. In Walker, R. G. and James, N. (eds) *Facies models and sea level change,* 3rd edn. Geological Association of Canada, St John's NF (A useful introduction to standard ichnofacies and application to sequence stratigraphy)

APPENDICES

Fieldwork basics

General preparation

- Base accommodation arranged
- Camping equipment and condition checked
- Repair outfits
- Water and food sources and supplies checked
- Water treatment equipment and tablets
- Cooking equipment in good order and fuel
- Clothing and footwear checked
  (suitable for climate and terrain)
- Personal insurance
- Check equipment in Box A.1

Box A.1 Field equipment

- Watch (check battery)
- Whistle
- Torch
- Hand lens (attached to string)
- Hammers
  - general use (2 lb, 1 kg)
  - with chisel end
  - toffee (4 oz, 100 g)
  - larger weights
- First-aid kit
- Water bottle
- Knife
- Survival blanket
- High-calorie food reserve
- Safety helmet
- Safety spectacles (eye goggles)
- Sunglasses
- Sunblock (suitable factor)
- Pickaxe, trenching tool
- Railroad pick
- Measuring tape
  - 2, 3 or 30 m length
  - metre stick
- Sieves
  - 10 mm goes above 1 mm
  - +0.6 mm at top (seawater)
  - +0.6 mm at top (dirty water)

- Grain size/sorting comparator
- Quadrat (or pegs, string)
- Notebook(s), pencils, coloured pens
- Water-resistant pens, wax pencil
- Correcting fluid
- Adhesive tape (waterproof)
- Carpet binding tape
- Logging sheets
  - general and specific
- Electronic notebook
- Telephone (if in service area)
- Cold chisels (with handle or glove)
  - general use (half-inch, 1 cm)
  - bolster chisel (brightly painted)
  - pinvices
- Long chisel (12 in, 30 cm) for shales
- Pointing trowel
- Pincers (to reduce slabs of shale)
- Paintbrush or hand brush for dusting
- Paintbrush for damping slabbed core
- Bucket or bowl

- Dilute hydrochloric acid (10%)
  - store in plastic dropper bottle
- Acidified stain (Alizarin red S)
- Packaging
  - newspaper, tissue paper *not* cotton wool
  - plastic tubes
  - bags in various sizes
- Glue (soluble), cotton bandages, scissors
- Camera(s) and accessory equipment
- Film, batteries
- Dental alginate for casting
- Releasing agents
  - wash-up liquid
  - hair shampoo
- Polystyrene for peels

For core shed
- Spray, sponge, paintbrush
- Illuminated reading glass (×2)
- Camera with flash and mounts

Location and access

- Topographical and geological maps
- Aerial photographs
- Local access (tide times and vegetation cover)
- Indemnity letters required?
- Permission to enter land and quarries obtained
- Indemnities signed and delivered

Has permission been obtained to collect and (for many countries) will it be possible to keep the material collected, and take it back? Several countries now have stringent rules relating to fossils, and to rocks in general. In Britain hammering and collecting is not allowed at many Sites of Special Scientific Interest (SSSIs). In North America and Australia pay close attention to regulations governing the taking of any geological material in National Parks, land administered by the Bureau of Land Management and other protected areas.

Some tips for the field

Sieving   Sieving is useful for concentrating teeth and small bones from loose sandy sediment. Do a trial run and examine what has passed through in a bucket or bowl. Do not overload the sieve or attempt to sieve mud or clay in the field.

Photographs   Take a photograph of a specimen before extraction, especially if it is likely to break up. Photograph trace fossils that are not extractable. With fossils of low relief it may be necessary to use a particular time of day. Support loose material at an angle similar to the inclination of the sun. Wetting (water, syrup) brings out bedding and lamination in terrigenous sediments. Always use a proper scale, never coins, hammer, lens cap, etc. A blob of reusable adhesive (e.g. Blu-tack), and a partly opened paper clip will fix the scale to any surface. Never think that the photograph will dispense with the need for written description and sketches.

New page   Start each day's notes on a separate page. Write the date and indicate your objectives; outline the general plan for the day. Organize the notes, even if this means jotting down some preliminary notes on a scrap of paper (or on centre pages to be removed later). Many notes are of a 'detective' type, or ideas to be sorted and analysed later.

Sketches   Use sketches liberally, supported by a photographic record; note the juxtaposition of blocks, or a broken specimen before removal. Record the location of each sketch and include a scale. Use squared paper as an aid or mark the face into squares. For small areas (paper sizes A3 to A5) place two rulers at right angles on the rock to aid in positioning features. Use simple, not feathery lines. Remember to record the trend of any faces sketched or photographed.

Folklore   There is a lot of folklore about collecting. Try to appreciate where the fossils are likely to be found. Also search tip heaps, scree slopes, the end of the conveyor belt, or along the shore (around low-water mark) where waves have concentrated pyritic fossils, or between boulders, or where rain and wash have concentrated small bones or teeth along gullies. Also examine walls, ploughed fields and immediately under the soil, especially under fallen trees. Be prepared to get down on your hands and knees. Expect the rock and fossil content to change at the boundaries of a quarry or section, including the floor.

Hammer carefully   Blows directed onto hard, massive limestone are rather ineffective. Search around for blocks that are suitable. For heavy blows, use the squared end of the hammer, full face; direct it towards yourself and wear goggles (especially for siliceous rock, chert or flint). Making outward, glancing blows generally leads to flakes of 'shrapnel' flying off the hammer. When using a chisel to work out the sediment around a fossil, direct the chisel away from the specimen. **Never** hit one hammer with another. When splitting rocks, use a substantial 'anvil'. Tap along the lamination with the square head of the hammer. To remove a specimen from a large slab, first chisel out a 'moat' around the specimen. Use tools appropriate to the job.

Preservation   When collecting fossil plants, pay particular attention to the mode of preservation; try to obtain material that shows the full organ and, if possible, organ relationships. Shales and mudstones with plant fossils are often deeply weathered and it is necessary to dig back to fresh material. When collecting from decalcified sandstones, only split material that is dry. Examine the edges for an indication of the frequency of mouldic preservation.

Labelling Label specimens before wrapping and add similar information to the wrapping paper or bag; enter the data in your field notebook. Mark specimens with way-up and, if necessary, strike and dip. Ring around small and less obvious specimens with a wax pencil. Use permanent marker or scratch temporary numbers. A blob of correcting fluid makes a good base, but makes specimens unsuitable for museum curation. Never place part and counterpart into direct contact, and keep thin slabs upright during transport.

Field preparation

For field preparation of material, the general rule to follow is to carry out the minimum of mechanical preparation. With blocks of fossiliferous muddy sediment, collected for later washing, try to prevent drying out by wrapping in damp newspaper and using plastic bags (preferably double). Spray 'timber' with a fungicide solution. Friable specimens may need supporting on a piece of wood or plastic. Spray fossil leaves with a soluble varnish (e.g. Tricolac) to reduce fragmentation. As a general rule, let covered specimens dry first. Most of the mud will then flake off. With specimens that have broken during extraction, sketch and photograph the distribution of the pieces, mark each and cross-tick adjacent edges, then pack for laboratory reassembling. If an attempt to mend is made, use a soluble glue such as Uhu and a piece of cotton bandage. Plasticine is useful for taking pulls of small specimens but silicone rubber is better, especially if there are parts of the mould that curve away into the rock, e.g. spines of a brachiopod. Latex in an ammonia solution (take care) is useful for casting larger surfaces of low relief. Greater penetration can be obtained if the specimen is moistened with a water/detergent solution; the first few coats should be of solution diluted with a detergent solution. Allow each coat to dry thoroughly. There are two main methods for making peels of loose sediment. Cellulose acetate and gauze can deal with fine-grained sands and silt. But this method is very time-consuming and a warm day is needed. (The fumes of acetone are obnoxious and poisonous.) Polyurethene foam is now replacing cellulose acetate (Skipper *et al.*, 1998). It is a quick, simple and lightweight method.

Plaster of Paris (heavy) can be used for support but do not apply it directly to a fossil, unless the surface is covered with waxed paper or aluminium foil. Reinforce with bandages. For displaying bioturbation in white chalk, spray with a solution of methylene blue, or gently smear light oil on a prepared smooth surface. Material that has been exposed to salt spray will have to be well soaked in freshwater. If some preliminary cleaning is attempted, e.g. removing caked and dried mud, take care not to lose important evidence such as the muddy fills to burrows, or the mud below a hardground crevice. For methods of impregnation and consolidation, refer to Rixon (1976) and Dowman (1970).

New techniques are published at intervals in the *Journal of Sedimentary Research*.

References

Dowman, G. A. (1970) *Conservation in field archaeology*. Methuen, London

Rixon, A. E. (1976) *Fossil animal remains: their preparation and conservation*. Athlone Press, London

Skipper, J., Ward, D. and Johnson, R. (1998) A rapid lightweight sediment peel technique using polyurethane foam. *Journal of Sedimentary Research* **68**: 516–18

Further reading

Bouma, A. (1969) *Methods for the study of sedimentary structures*. Kreiger, Huntingdon NY (Reprinted 1979 by J. Wiley, New York)

Brunton, C. H. C., Besterman, T. P. and Cooper, J. A. (1985) *Guidelines for the curation of palaeontological materials*. Miscellaneous Paper 17, Geological Society, London

Collins, C. (ed.) (1995) *The care and conservation of palaeontological material*. Butterworth-Heinemann, Oxford

Holme, N. A. and McIntyre, A. D. (eds) (1984) *Methods for the study of marine benthos*. Blackwell, Oxford

Kummel, B. and Raup, D. (eds) (1965) *Handbook of paleontological techniques*. W. H. Freeman, New York (Mainly laboratory techniques)

Tucker, M. E. (1989) *Techniques in sedimentology*. Blackwell Science, Oxford

Wolberg, D. and Reinard, P. (1997) *Collecting the natural world: legal requirements and personal liability for collecting plants, animals, rocks, minerals and fossils*. Geoscience Press, Tuscon AZ

Field statistics

At many fossiliferous sites it is likely that a number of questions will arise requiring statistical analysis:

- Is there a preferred orientation to............?
- Is the dispersion of..........clustered?
- Do the encrusting organisms show a substrate preference?
- Are these two species really distinct?
- How similar are these two samples?

A pocket calculator will be required. More sophisticated analysis can be carried out in the laboratory with the aid of software such as PALSTAT (Ryan *et al.*, 1994).

B.1 Dispersion

Sessile plants and animals may or may not be influenced by the presence of other members of the same species when they become distributed early in life. If they are, then the resulting distribution of individuals will be either 'regular' or 'clumped' (Fig. 5.22), depending on whether they avoid each other or seek each other out. No influence of any individual on any other will result in a random distribution. Measurements have to be made in the field. The nearest-neighbour technique is the one to use:

1 Measure the distance r (to the nearest mm or cm) between the centre of each individual (or colony) and its nearest neighbour. Lightly mark each sample (but not its neighbour) as it is measured to avoid remeasurement.
2 Calculate the mean M of the observed distances

$$M = \frac{\Sigma r}{n}$$

where n is the number of measurements.
3 Calculate the density D of individuals over the sampled area (in mm^2 cm^2)

$$D = \frac{n}{\text{area}}$$

4 Calculate the expected mean distance E between neighbours

$$E = \frac{1}{2\sqrt{D}}$$

5 The ratio $R = M/E$ is a measure of the degree to which the observed distances depart from random. A value of $R = 1$ indicates randomness, $R = 0$ indicates maximum aggregation (all fossils at one point), and $R = 2.15$ indicates an even distribution.
6 To test the significance of the departure of the mean observed distance from the expected mean distance, calculate the standard variate

$$c = \frac{M - E}{0.261\,36/\sqrt{nD}}$$

For 95% probability ($p = 0.05$) $c = 1.96$
For 99% probability ($p = 0.01$) $c = 2.58$

Note   Measurements must be taken on individual bedding surfaces only. If the populations are large, take randomly selected individuals, but be sure to measure the distance from each selected sample to the actual nearest neighbour in the whole population. In Fig. B.1 the bryozoans are clumped. The *Skolithos*

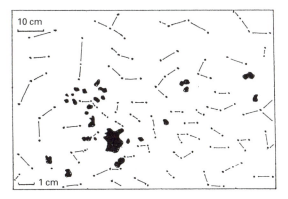

Fig. B.1 Black blobs (10 cm scale) show the distribution of encrusting lamellate bryozoa on part of an Ordovician hardground analysed by Palmer and Palmer (1977); other fauna are omitted for simplification. Dots (1 cm scale) map part of a quadrat with *Skolithos* burrows (Lower Cambrian, Labrador), and nearest neighbours are connected by lines; after Pemberton and Frey (1984).

burrows show $R = 1.65$, indicating a significant departure from the random towards a regular spacing, probably associated with suspension feeding.

B.2 Variation

The eye is particularly good at distinguishing variation and may tend to overestimate diversity. Two species which each show considerable variation, are they really distinct? Although calipers are better, a transparent ruler can be used (inefficiently) to measure parameters, e.g. length (l), width (w) and height (h) of a rhynchonellid brachiopod. Each parameter can be plotted as a histogram. Unimodal plots will suggest that only one species is present. A ratio (w/h) can be plotted against another parameter (l) on a graph. This may distinguish two species as separate clusters of points. Alternatively, plot points for each species (the eye has recognized) and plot the regression lines (reduced major axis) for each:

Plot $\quad Y = a + bx \quad$ with $\quad a = \bar{Y} - b\bar{X}$
$$b = \Sigma XY / \Sigma X^2$$
where $\quad \bar{Y}$ = mean of Y values
$\quad\quad\quad \bar{X}$ = mean of X values

Are the regression lines significantly different?

B.3 Orientation analysis

There are three main patterns of preferred orientation in a plane: unimodal, a trend (bimodal) and polymodal.

Measure the trend of the most accurately measured parameter, e.g. length of crinoid stems, width of a brachiopod (on a 1–180° scale); and measure the direction, e.g. belemnite apex, length of a brachiopod away from umbo (on a 1–360° scale), as shown in Fig. B.2. A circular histogram may be plotted (selecting appropriate class intervals, 10°, 15°, 20° or 30°, depending on the amount of data) by sector area, sector radius or as a spoke proportional to the class frequency. Note that frequencies drawn by sector radii and 'spokes' result in distortion. The radius of each sector is given by $r^2 = 2Af/Nw$, where r is the sector radius, A is the total area of the histogram, f is the class frequency, N is the sample size and w is the class width in radians (1 rad = $180/\pi$ degrees).

Alternatively, plot the direction as a circular plot, or trend as a circular plot having doubled the angles (so that north will then be 0°, 180° and south 90°, 270°). The circular median direction can then be readily found by searching for the direction that aims for that section of the circumference with the greatest point density, and which divides the sample into two equal parts.

Confidence tests can be carried out later, or see Cheeney (1983) for the Rayleigh test. Where the number of observations is low, and the pattern unclear, use the chi-square test (Section B.4) to see whether the data items are clumped, compared with the expected even distribution.

On rock faces, elongate and cylindrical objects, such as crinoid stems or tall-spired gastropods, are often viewed end on. There may appear to be a clear preferred trend, but orientations above 50° (to the face) give ellipses deceptively close to circular.

B.4 Chi-square test (Lamboy and Lesnikowska, 1988)

- Are the observed frequencies of valve sorting between samples from two beds, A and B, significantly different? Bed A is a bioturbated shelly muddy sandstone, and bed B is a clearly allochthonous, shelly, well-sorted, fine-grained sandstone.

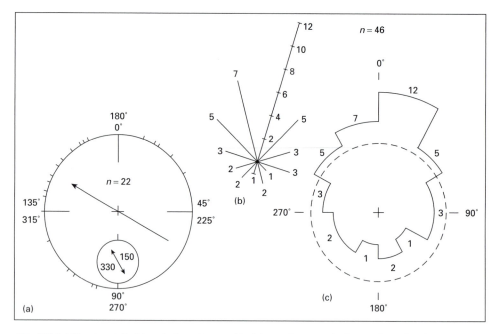

Fig. B.2 (a) Circular plot of trends for a linear object (e.g. plant stem); the median direction is shown bisecting the sample. Here the angles have been doubled, so the median trend is 150° (*n* = 22). (b) Spoke histogram of the same data as in part (c); the data is spaced in 30° classes and plotted with radius proportional to the class frequency. (c) Circular histogram for 46 measurements grouped in 30° classes and plotted with sector area proportional to class frequency. To simplify plotting, make the total area of histogram equal to unity (*A* = 1), so that for a class frequency *f* = 5 then $r^2 = (2 \times 1 \times 5)/(46 \times 0.523)$.

Observed frequencies

	Pedicle valves	Brachial valves	Total
Bed A	198	308	506
Bed B	196	258	454
	394	566	960

Expected frequencies

	Pedicle valves	Brachial valves	Total
Bed A	$\dfrac{394}{960} \times 506 = 208$	298[a]	506
Bed B	$\dfrac{394}{960} \times 506 = 186$	286	454
	394	566	960

[a] By subtraction: 506 − 208 = 298.

$$\chi^2 = \sum \frac{(\text{observed} - \text{expected})^2}{\text{expected}}$$

i.e. $\dfrac{(198-208)^2}{208} + \dfrac{(308-298)^2}{298} + \dfrac{(196-186)^2}{186} + \dfrac{(258-268)^2}{268} = 1.6$

which is not significant at $p = 0.01$ (χ^2 with one degree of freedom is 6.63) or at $p = 0.1$ ($\chi^2 = 2.71$).

- Is the amount of breakage between two samples significant? The expected frequencies for the null hypothesis that there is no difference, can be estimated from the combined proportion of broken to whole shells in each sample.
- Is the proportion of reclining to pedunculate shells in the samples significantly different?

Substrate preference

Substrate preference can also be assessed by using the chi-square test to ascertain the probability that one substrate was preferred to another, e.g. by encrusters.

1 Make a general inspection of the substrates and encrusters to identify the types of each present.

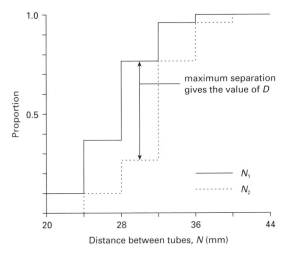

Fig. B.3 Cumulative histogram for measurements on two populations of U-burrows.

2 Taking each substrate (of similar area) in turn, record the number O of individuals of each different encrusting species.
3 Calculate for each encrusting species, the expected number E of encrusters on each substrate:

$$E = \frac{\text{total number of substrates of each type}}{\text{total number of each encrusting species}}$$

$$\chi^2 = \sum \frac{(\text{observed} - \text{expected})^2}{\text{expected}}$$

B.5 A test for small samples (less than 40)

We are often faced with the problem of trying to discriminate between small samples. For instance, might two populations of less than 40 individuals belong to the same taxon or a different taxon? Measurements of the distances between the tubes of U-burrows on two bedding surfaces were made for 20 pairs on each surface. These measurements are set out on a cumulative histogram (Fig. B.3); where N_1 is the distance between the tubes on surface 1 and N_2 is the distance on surface 2. The null hypothesis is that the two samples are of the same taxon and that differences are due to factors such as growth and the variation within each population. The Kolmogorov–Smirnov test statistic D for a 95% probability that the two samples are of the same ichnotaxon is given by

$$1.36\sqrt{(N_1 + N_2)/(N_1 \times N_2)}$$

and in this case it is

$$1.36\sqrt{(20 + 20)/(20 \times 20)} = 0.43$$

In Fig. B.3 the maximum separation (observed value of D) is 0.50, in excess of the expected value. It is thus reasonable to infer that two ichnotaxa are present. If the observed separation had been less, then the null hypothesis could have been supported.

For $p = 0.01$ (99% probability) D is given by

$$1.63\sqrt{(N_1 + N_2)/(N_1 \times N_2)}$$

B.6 Sampling and rarefaction curves

A sampling curve was mentioned in Section 2.5. It may sometimes be useful to generate a rarefaction curve, which might also be called an artificial sampling curve. From a single sample of determined diversity, the curve predicts the expected diversity for *smaller* samples. Extending the curve can also give an indication of the way diversity might increase in *larger* samples. Sanders' object, when he introduced the curve, was to demonstrate the marked differences in curve shape that are found for different environments. Consider this example.

A sample of 200 individuals contains 20 species, arranged with the following abundances:

Species	n	Sample %	Cumulative %
1	100	50	50
2	40	20	70
3	24	12	82
4	8	4	86
5	6	3	89
6	4	2	91
7–10	2 each	1% each	95
11–20	1 each	0.5% each	100

Species	Samples	
	A	B
1	20	50
2	20	35
3	20	10
4	20	4
5	20	1
	$e = 1.0$	0.7

For a sample of 100 individuals, one might expect 15 species. Each specimen represents 1% of the sample. From the tables above there are 10 species that each represent 1% or more, and they account for 95%. One would not expect to find more than 5 more species (100 − 95). For a sample of 50 individuals, one might expect to find 11 species. Each specimen represents 2% of the sample. From the tables above, there are 6 species that each represent 2% or more, and these 6 species account for 91%. The residual 91% cannot be expected to yield more than 5 species. The expected number of species is 6 + 5 = 11.

B.7 Diversity and species frequency

Is there a significant difference in the way the species are distributed through two samples (collections)? How even are their diversities? Although the diversity of two samples may be identical, say with 10 species in each, in one sample each species may be present in equal numbers, whereas in the other sample two species may account for 80% of the individuals. The two samples may be compared by the evenness index:

$$e = \frac{\Sigma(n_i/n) \log (n_i/n)}{\log S}$$

where S = number of species
n_i = total number of individuals in the ith species
n = total number of individuals

An index that combines richness and evenness (Section 5.3.8) is the Shannon index:

$$(H) - \Sigma(n_i/n) \log (n_i/n)$$

References

Cheeney, R. F. (1983) *Statistical methods in geology*. George Allen & Unwin, London

Lamboy, W. and Lesnikowska, A. (1988) Some statistical methods useful in the analysis of plant paleoecological data. *Palaios* **3**: 86–94

Palmer, T. J. and Palmer, C. D. (1977) Faunal distribution and colonization strategy in a Middle Ordovician hardground community. *Lethaia* **10**: 179–99

Pemberton, S. G. and Frey, R. W. (1984) Quantitative methods in ichnology: spatial distribution among populations. *Lethaia* **17**: 33–49

Ryan, P. D., Harper, D. A. T. and Whalley, J. S. (1994) *PALSTAT* Chapman & Hall, London

Further reading

Jones, B. (1988) Biostatistics in paleobiology. *Geoscience Canada* **15**: 3–22

Walden, A. T. *et al.* (1992) *Statistics in the environmental and earth sciences*. Edward Arnold and Halstead Press, London

Wratten, S. D. and Fry, G. L. A. (1980) *Field and laboratory exercises in ecology* Edward Arnold, London

APPENDIX C — Biomineralization

Trophism	Depth	Salinity	Dominant and (minor) organisms	Mineralogy	Trophism	Depth	Salinity	Dominant and (minor) organisms	Mineralogy
			Algae			SD	N	*Graptolithina*	O
autotrophs	S (3)	N	green	A	Su	S(SL) {	N	*Brachiopoda*	C (M)
		F (25)	charophytes	C (25)			N–T	most inarticulates	P
	S (26)	N–T	red	M (A)				*Molluscs*	
		N	coccoliths	C					
		F–H	diatoms	S	W	S(SD)	F–H	Gastropoda	A (4)
	S	F–H	skeletal	C, A	Dp	S	N	Scaphopoda	A
			stromatolites		Cr (17)	S–D (17)	N	Cephalopoda	A (5)
			(blue-green algae)					Bivalvia	(24)
		F (H?)	calcispheres (Palaeozoic)	M?	Dp	S	N	protobranchs	A
								oysters	C (6)
		N	calcispheres (Mesozoic)	M				mytilids	B (7)
					Su (18, 1)			pinnids	B (8)
			Foraminifera					pterids	B (8)
								pectens	C (9)
(21) {	SD	B–T	Textulariina	G (14)	Su (18, 1)	S	(19) {	anomids	C (10)
(22) {	S	N–H	Miliolina	M				limids	B (11)
(22)	SSL	N	Rotalina	CMA (15)				hippuritids	B
	S	N	Fusulinina	C				(rudists)	
								Chama	A (12)
(21)		N	*Radiolaria*	S				rest	A
Su	S	N	*Archaeocyatha*	Mg	W	SD	F–H	*Annelids*	OPG
			Sponges		SU		F–T	serpulids	M (A)
	SD	N(B)	Demospongea	S, O (3)				*Arthropods*	
Su {	S		Lithistids		(20)		N	trilobites	C
	SD	N	Hexactinellids	S	W	S	F–H	ostracodes	CM
	S		Calcarea	CMA	Su		T	cirripedes	C
			Sclerospongea	AC (23)	W		F–H	decapods	CP (13)
			Stromatoporoidea		Su		F(B/N)	branchiopods	CP
Su?	S	B?–T {	Mesozoic	A	W			insects, arachnids	O
			Palaeozoic	M (2)	Cr	S	F–B	merostomes	O
			Coelenterates					*Echinoderms*	
			Hydrozoa	A	Su			stemmed (28)	
			Scleractinia (28)	A	Dp/Bw(Su)	S/SL	N {	echinoids	M
			Octocorallia	MA	Cr			'starfish'	
Mc	S (16)	N {	Rugosa	C	Su/Cr			brittlestars	
			Tabulata (27)	C (A)				*Vertebrates*	
	SD		(conularids)	P				bone, teeth	P
			Bryozoans		W	–	–	eggshell	C
Su	S(Sl)	T {	Cheilostomata	M (A)				otoliths	A
			rest	C	?Sc/Cr	SSL?	N	conodonts	P

Trophism

Su	suspension	taking food from suspended matter in water ⎤ without need to break up particles
Dp	deposit	taking food from substrate ⎦
BW	browser	scraping plant material from surfaces or chewing vegetative plant matter (grades into deposit feeding)
Sc	scavenger	eats dead organisms (grades into deposit feeding)
Mc	microcarnivore	supplemented in hermatypic Scleractinia by photosymbionts
W	wide range	

Mineralogy	Salinity	Depth (marine benthos)
C calcite	N stenohaline	S shelf (to 200 m)
M Mg calcite:	normal marine	SL slope (to 2000 m)
>4 mol % MgCO₃	(30–40‰)	D deep
A aragonite	T euryhaline	SLL shelf to slope
B bimineralic C, A	greater	SD shelf to deep
S siliceous	tolerance	
O organic	(20–50‰)	
P phosphatic	B brackish	
G agglutinating	H hypersaline	
	F freshwater	

Notes

(1) *Solemya* chemosymbiont suppl.
(2) Affinities of some in dispute.
(3) Spongin or S + spongin, S readily replaced by C, typically ⪢50 m.
(4) Most extant archaeogastropods (*Pleurotomsria, Patella, Trochus,* neritids) with outer C, inner A.
(5) Aptychus, guard, rhyncholites C.
(6) A only in ligament and myostraca.
(7) Variable: entirely A or outer prismatic C and inner A (nacre).
(8) Outer C prisms, inner nacre A.
(9) C (foliated) (some A in middle layer).
(10) Outer C foliated, assd. with A (cross-laminar).
(11) Inner A (cross-laminar).
(12) Two species with C.
(13) C as reinforcement.
(14) Habitat-derived grains (quartz, sponge spicules, other foraminifera); cement organic, calcareous or FeO.
(15) Most larger M, and globeriginids C.
(16) Hermatypic, oligotrophic Scleractinia typically <20 m, 20–25 ˚C.
(17) Ammonites may have had wide range of depth and trophism.
(18) Mainly Su except Dp *Tellina* and related genera; *Teredo* wood.
(19) Marine except for F: unionoids and M–B–F mytilacids.
(20) Probably W.
(21) Pseudopodial trapping.
(22) Some supplemented by commensal autotrophs (zooxanthellae).
(23) Plus spicules S.
(24) C only in some epifaunal groups (or in groups with epifaunal ancestry); tube lining of *Teredo* C, others A or?
(25) M in brackish environments; some modern forms tolerate hypersalinity.
(26) Mostly shallow, but to 100 m and greater; articulated intertidal.
(27) A in Tetradiidae.
(28) Stenoxic.

Principal source: Lowenstam, H. H. and Weiner, S. (1989) *On Biomineralization*, Oxford University Press, Oxford

Checklists

OUTCROP

- Location
- Dip/strike
- Lithology
  colour, grain size/
  sorting
- Bedding (cm–m)
- Cleavage(s), jointing
- Strength, hardness

AUTOCHTHONY

- Natural orientation
- Skeletal dissociation (L, R)
- Shell attitude
- Size sorting
- Bedding orientation
- Breakage/damage
- Lithology/matrix
- Clustering
- Adjacent sediment

ECOLOGICAL FACTORS

- Bathymetry
- Temperature
- Salinity
- Substrate
- Oxygenation
- Interrelationships
- Abundance
- Diversity
- Dispersion

For plants
- Temperature,
  precipitation
- Illumination
- Soil
- Biological
  interrelationships

BUILDUPS

- Autochthony
- Ecological role
  binding, encrusting,
    stabilizing
  frame builders,
    bafflers
  loose sediment
    producers
- Growth form
- Growth rate
- Growth stage
  domination,
    diversification
  colonization,
    stabilization
- Local relief
- Matrix
- Cavities
- Diagenesis

ALLOCHTHONOUS SKELETAL ACCUMULATIONS

Local (cm–m)
- Bedding stringers,
  pavements
- Current shadows, clusters
- Winnowed layers
- Gutter fills, burrow fills
- Desiccation structures
- Solution fissures
- Tectonic fissures

Extensive (m–km)
- Lags, relict
  accumulations
- Storm beds, tempestites,
  turbidites
- Pelagic accumulations

TRACE FOSSIL TAPHONOMY AND DESCRIPTION

- Mode of preservation
  epichnia, endichnia,
    exichnia
  hypichnia (convex,
    concave)
  undertrack, ichnoclast
- Ichnofabric
- Endogenic, exogenic
- Sediment consistency
- Ordering
- Tiering
- Variation in preservation
  mode
- Identification, description

BODY FOSSIL TAPHONOMY

- Mode of preservation
  original shell, skeleton
  replacement, encrustation
  external/internal
  impression
  composite mould,
    crystal core
  tectonic distortion
  compaction cast
- Skeletal dissociation
- Epizoan, epilith
- Preferred orientation
  gravity, light
    food, current

APPROACH TO BASIN ANALYSIS

- Lithology
- Bedding/stratinomy
- Facies analysis
- Facies sequence analysis
  recognition of key
    surfaces
- Basin analysis
- (Global analysis)

Glossary

Introduction

Useful figures

- Environments (Fig. 2.1)
- Bedding (Fig. 2.2)
- Coal types (Fig. 3.11)
- Trace fossils (Fig. 8.1)

References

Allaby, A. and Allaby, M. (1990) *The Concise Oxford Dictionary of Earth Sciences*. Oxford University Press, Oxford (6000 entries)

Jackson, J. A. (ed.) (1997) *Glossary of Geology*, 4th edn. American Geological Institute, Alexandria VA (over 37 000 terms)

The glossary

active fill   material actively passed into burrow by animal, often pelleted; equivalent to backfill

actuopalaeontology   the study of organisms and their taphonomy in modern environments in order to understand fossils better

aerobic   conditions are aerobic in the presence of free oxygen (>2.0 ml l^{-1} O$_2$)

allochthonous   a fossil is allochthonous if it has been transported from the environment in which it originally lived (cf. autochthonous)

allogenic   generated externally

allostratigraphy   subdivision of the rock record with bodies of rock (facies succession) on the basis of bounding discontinuities and correlative conformities; it is thus of potentially wider application than lithostratigraphy by emphasizing bounding surfaces

anactualistic (non-actualistic, non-uniformitarian)   not as (typically) observed in modern sedimentary environments

anoxic   devoid of oxygen

assemblage   an accumulation of dead material of any age

association   a consistently recurring group of fossils that may be assumed to have lived together, but which is only part of a total palaeocommunity

astogeny   development by asexual reproduction leading to branched colonial organism, such as colonial (compound) corals, bryozoans and graptolites

astrate   lacking stratification

autochthonous   a fossil is autochthonous if it is found in the place where it originally lived (cf. allochthonous)

aut(o)ecology   the study of the (palaeo)ecology of an individual organism or species

autogenic   self-generated

autogenic succession   an ecological succession resulting from factors inherent in the community, self-generated

basin analysis   analysis of the extent and context of facies within a basin

benthonic   relating to the substrate

biocoenosis   an association of living organisms

biogenic sedimentary structure   *see* trace fossil

bioimmuration   encrustation of soft tissues by skeletal organism, to leave a natural impression of the soft tissue

biostratinomy   *see* stratinomy

bioturbation   process of sediment mixing by organisms and breakdown of primary sedimentary structures

body fossil   the remains or representation of a whole or part of an animal or plant, e.g. shell, impression of a jellyfish, or skin, charcoalified plant

boring   biogenic sedimentary structure excavated into a hard substrate by chemical means or by grinding

Bouma cycle   a predictable succession of lithologies, hydraulically related, that make up a turbidite bed

buildup   a body that has original topographic relief. In this text autochthonous buildup (Fig. 2.6) is used in a wide sense for an accumulation of dominantly autochthonous organisms; though generally evident, a framework (growth fabric) may be obscured by taphonomic processes

burrow   biogenic sedimentary structure in soft sediment below the sedimentary surface

census assemblage   a near to perfect record of the live (shelled) biota

chemosymbiont organism having commensal autolitho-trophic bacteria and thus a nutrient source

cleavage relief with trace fossils, refers to deviation of cleavage around coarser-grained burrow

climax community (of body or trace fossils) a feature of stable conditions allowing a high diversity with narrow ecological niches

coelobite cavity-dwelling organism (applied in palae-ontology, especially to encrusting and nestling organisms within borings)

colonization window period of time available for success-ful colonization of the substrate

commensalism a relationship between two organisms in which one benefits but is not injurious to the other

community group of animals and/or plants that live together; a palaeocommunity is a community of animals and/or plants that are assumed to have lived together in the past

composite trace fossil distinct ichnotaxon or ichnotaxa within another ichnotaxon

compound trace fossil one ichnotaxon passing into another ichnotaxon

condensed assemblage skeletal assemblage with admix-ture of shells of significantly greater age

coquina detrital limestone composed chiefly or wholly of mechanically transported skeletal material

crypt a cavity

cryptic refers to concealed, cavity dwelling of organisms

cycle sedimentary sequence reflecting recurrent, repeated events, so that conditions at the beginning and the end are the same

derived fossil skeletal material derived (reworked) from an earlier sedimentary cycle

diagenesis the changes (chemical, physical and biological) that occur in a sediment after its initial deposition, and during and after lithification, but excluding weathering and metamorphism

diastem (non-sequence) a relatively short interruption in sedimentation

disconformity unconformity without structural discordance

disjunct where a taxon exists in separate areas

dysaerobic very little free oxygen (0.2–2.0 ml l^{-1} O_2)

ecology the study of the interactions of organisms with one another and with the environment

edaphic relating to soil

elite trace fossil ichnotaxon that is especially conspicu-ous because of diagenetic enhancement

endemic restricted geographically

energetics (ecological energetics) energy transformations within ecosystems (community and environment)

epifauna (a) fauna living upon rather than below surface of the seafloor; (b) fauna living attached to rocks, etc.

epilith any organism attached to rock or a bioclast

epizoan any (animal) organism attached to a living organ-ism; *adj.* epizoic; *syn.* epizoite

euryhaline organisms tolerant of a wide range of salinity; salinity range 30–40‰

eustacy (or eustasy) refers to global sea-level change

eutrophic said of body of water with high level of plant nutrients

exaerobic biofacies with epibenthic chemisymbionts

exotic organism that has been introduced

fabric spatial arrangement of particles in a sediment

facies unit of rock defined by its total geological character (geometry, lithology, sedimentary features and fossil con-tent); this is the way it is used here – in an observational and descriptive sense – but it is also often used in a genetic sense, e.g. turbidite facies, or in an environ-mental sense, e.g. shallow-marine facies, or in a tectono-sedimentary sense, e.g. post-orogenic facies

facies sequence analysis analysis of the vertical organiza-tion of facies

firmground stiff but uncemented substrate

formation the formation is the primary local litho-stratigraphical unit; it is mappable and should possess internal homogeneity with distinctive lithological features; thickness should not be a determining feature

form genus a taxon of convenience used in the classifica-tion of fossils of problematic relationship; the taxon may be unassignable to a higher taxonomic category

fossil any remains, trace or imprint of a plant or animal that has been preserved, by natural processes, in the Earth's crust since some past geologic or prehistoric time

Fossil-Lagerstätte *see* fossil-ore

fossil-ore (Fossil-Lagerstätte) any fossiliferous site yield-ing an unusual amount of palaeobiological information because of the excellence of preservation (e.g. soft tissue) or because of the abundance of material; an obsolete definition is fossiliferous iron ore

gallery (of trace fossil) burrow more or less parallel with substrate surface opening from shaft

guild a group of species that exploit the same class of environmental resources in a similar way, without regard to taxonomic position, e.g. construction guild of reefs

hardground a sedimentary non-sequence during which the substrate became lithified; typically recognized by encrusting or boring organisms

heterochrony different rates of reaching maturity, neoteny, acceleration; change in timing of ontogenetic events

homoeomorphy superficial resemblance between taxa, but dissimilarity in detail; biologists define it more narrowly, relating it only to species

hummocky cross-stratification type of cross-stratification associated with laminae that dip at low angles and with little or no preferred orientation; sole generally planar but succeeding bounding surface hummocky; associated with storm events

hypautochthonous refers to accumulation of plant fossils that are more or less in the general area in which they grew, e.g. peat swamp

hypersaline denoting salinity substantially greater than that of normal sea water, >40‰

ichnocoenosis trace (fossils) formed by a coexisting benthic community

ichnofabric   sediment texture and internal structure due to bioturbation and bioerosion; the ichnofabric integrates the taphonomic and stratinomic aspects of the trace fossils with the primary fabric

ichnofabric constituent diagram   representation of primary stratification, ordering and tiering

ichnofacies   recurring associations of trace fossils on all scales

ichnofossil   *see* trace fossil

ichnoguild   a group of trace fossils that expresses a similar sort of behaviour and belongs to the same group and occupies a similar tier level

indigenous assemblage   production by the local fauna

inertinite   macerals characterized by high carbon content

infauna   aquatic organisms that live essentially below the substrate surface

in situ   in place in the rock at outcrop (or core), as distinct from in scree, soil or clitter; also applies to spores within sporangia, seeds within pod, etc.

kerogen   acid-insoluble organic residual

lazarus taxon   reappearance of a taxon after apparent extinction

lebensspur (*plural* lebensspuren)   *see* trace fossil

lumachelle   an accumulation of shells, especially oysters; *see* coquina

maceral   microscopic constituents of coals

margin (burrow)   the regular or irregular boundary of a burrow. It may be unlined, or have a lining that is smooth or pelleted (±faecal), thin or thick and has been actively or passively formed; it may also be impregnated and deformed

member   a part of a formation recognized by a particular lithological peculiarity; it is not necessarily mappable

meniscate   active burrow fill of thin, dish-shaped laminae formed behind advancing animal

mesohaline   salinity range 18–30‰

meteoric   water of recent atmospheric origin

mutualism   a relationship beneficial to both organisms

necrotic processes   processes of death

nektonic   swimming

neomorphic   skeletal replacement by another mineral (or the same mineral) without an intervening void stage

neoteny   acceleration of sexual maturity with retention of many juvenile characters; *syn.* paedogenesis

obrution   smothering, rapid burial

oligohaline   salinity range 0–5‰

oligotrophic   said of water body with deficiency of plant nutrients

ontogeny   development of an individual organism; *see also* astogeny

opportunists   said of organisms able to colonize vacant niches rapidly

ordering   sequence of burrows/borings in a bed recognized by cross-cutting relationships; burrows relate to a single substrate surface; absence of ordering indicates contemporaneous colonization

palaeocommunity   *see* community

palaeoecology   the study of the interactions of organisms with one another and with the environment in the geological past

palaeosol   a fossil soil; most fossil soils do not show roots, but they do provide indication of modification of the material by near-surface (terrestrial) organic, chemical and physical processes

palimpsest (for trace fossils)   superimposition of two ichnofacies

palynofacies (analysis)   the quantitative analysis of the environmentally typical residue of a sediment, after elimination of the mineral phase by acid digestion; the residue comprises spores, pollen, foraminiferal test linings, microplankton, plant and animal cuticle, inertinite, vitrinite and amorphous organic matter

parataxonomy   system of classification in which parts of an animal or plant (or work) are classified separately; *see* form genus

parautochthonous   a fossil is parautochthonous if only little post-mortal transport has occurred and burial is, usually, in the same environment (biotope)

pelagic   swimming or floating organisms

phenotypic (variation)   variation (in morphology) due to environmental factors, e.g. crowding

phobotaxis   behavioural avoidance of an earlier formed burrow (by same or other organisms) through stopping short, running parallel or overcrossing; achieved by chemo- or physicosensory methods

phreatic   water in saturated zone; equivalent to groundwater

phyletic gradualism   evolutionary process or pattern in which morphology changes gradually and continuously

placer   mechanical concentration

planktonic   floating

poikiloaerobic   variable oxygenation state with temporary lack of oxygen

polyhaline   salinity range 18–30‰

prefossilization   diagenetic change to fossil material in the primary sedimentary cycle, before final burial

pseudoplankton   attached to driftwood or drifting empty shells of cephalopods

punctuated equilibrium   evolutionary process or pattern in which morphologic change is concentrated within a relatively short interval of a species range; a period of stasis follows

reef   a term which seems to defy a definition that is useful to the field geologist; use the term *buildup* and justify it

regression   a change that brings non-marine, nearshore or shallow-water conditions over deeper offshore conditions, an extension of land areas

remanié   shells reworked from appreciably older deposit

rheotaxis   (preferred) orientation due to stream flow

rhythm   term used interchangeably with cycle

riparian   living beside rivers or streams

sapropel   plant material degraded under anaerobic conditions

sequence conformable succession of facies arranged in a predictable manner; generally bounded by sharp junction (boundary)

sequence stratigraphy the application of seismic stratigraphic interpretation techniques to sedimentary basin analysis

shaft (of trace fossil) burrow usually normal to stratification surface

Signor-Lipps effect apparent gradual decline (or abrupt extinction) in taxa due to sampling errors or stratigraphic/environmental effects, especially of low-frequency taxa, whereas the actual situation is mass extinction (or gradual decline)

spreite (*plural* spreiten) the successive margins of a displaced burrow, produced as animal shifts broadside through sediment

stenohaline organisms tolerant to a narrow range of salinity

stratinomy the processes between death of an organism and its final burial; also applies to trace fossils, e.g. at hardgrounds

synecology study of the relationship between (palaeo)-communities and their environments

taphocoenosis an assemblage of fossils characterized by similar modes of preservation, and normally, similar stratinomic history

taphofacies suites of sedimentary rock characterized by particular preservational (taphonomic) features

taphonomic feedback change in community structure due to changes in the nature of the substrate, associated with dead skeletal material and bioturbation

taphonomy all the changes that occur to an organism (or its work) between death and discovery as a fossil (Fig. 3.1)

taxonomy the theory and practice of classifying animals and plants

tempestite the sedimentary product of a tempest, typically displaying a sequence of sedimentary depositional structures attributable to storm flow and wane, and involving appreciable lateral transport; the base is typically erosional and the top gradational and/or bioturbated

texture relative proportions of particles, especially particles in different size classes; the physical appearance of a rock

thanatocoenosis a burial assemblage, where each fossil is in the position in which it died (autochthonous)

tiering trace fossils and body fossils are tiered above and below the substrate surface; tiering is a synonym of the biological term stratification

time averaging mixing of skeletal components by physical or biological processes over a relatively short time span, thereby averaging out temporary, stochastic (random) fluctuations within a community

toponomy mode of preservation of biogenic sedimentary structures

trace fossil evidence of the activity of an organism in ancient sediments, e.g. fossil burrows and footprints; also includes fossil faecal pellets and coprolites

track impression on soft to firm surface made by foot or equivalent

transgression a change that brings deeper-water conditions over shallower-water conditions, or a spread of sea over land areas

trophism nutrition involving metabolic exchange in the tissues

tube burrow with prominent lining, e.g. tube worm burrow

turbidite the product of a turbidity current, typically demonstrating a sequence of sedimentary depositional structures associated with a decelerating current; erosion is typical at the base; the top is typically gradational with or without bioturbation

unconformity a substantial stratigraphical break in the rock record, with or without structural discordance; *see also* diastem

vadose water of aerobic zone

wall (of trace fossil) term best avoided as refers to both burrow margin and lining

xenomorph growth of encrusting organism follows (and hence reveals) encrusted organism

Index